FUNDAMENTALS OF MOLECULAR PHYSICS

FUNDAMENTALS OF MOLECULAR PHYSICS

Josef Carter

Oxford Book Company
Jaipur ▪ New Delhi

Published By

Oxford Book Company

Regd. Off. "Bony Residency", Gate No. 2
Opp. Tilak Public School, Vishweswariya Nagar,
Gopalpura Road, Jaipur - 302018, India
Ph: 0141 - 2761280, Telefax: 2761381
email - oxfordbookjpr@gmail.com
website - oxfordbookcompany.com

Delhi Office: 102, 1st Floor, Satyam House,
4327/3, Ansari Road, Darya Ganj,
New Delhi - 110002, India
Ph: 011-45652440

Distributed by
KEYA INTERNATIONAL PRESS LLP
E-mail: keyainternationalpressllp@gmail.com

First published in India, 2024

ISBN : 9789355240873

Typeset by Shivangi Computers, Jaipur

Printed and bound in India by Replika Press Pvt. Ltd.

Contents

Preface

Molecular physics is the study of the physical properties of molecules, the chemical bonds between atoms as well as the molecular dynamics. Its most important experimental techniques are the various types of spectroscopy. The field is closely related to atomic physics and overlaps greatly with theoretical chemistry, physical chemistry and chemical physics. Additionally to the electronic excitation states which are known from atoms, molecules are able to rotate and to vibrate. These rotations and vibrations are quantized, there are discrete energy levels. The smallest energy differences exist between different rotational states, therefore pure rotational spectra are in the far infrared region (about 30 - 150 μm wavelength) of the electromagnetic spectrum. Vibrational spectra are in the near infrared (about 1 - 5 μm) and spectra resulting from electronic transitions are mostly in the visible and ultraviolet regions. From measuring rotational and vibrational spectra properties of molecules like the distance between the nuclei can be calculated. One important aspect of molecular physics is that the essential atomic orbital theory in the field of atomic physics expands to the molecular orbital theory. Molecular modelling encompasses all theoretical methods and computational techniques used to model or mimic the behaviour of molecules. The techniques are used in the fields of computational chemistry, drug design, computational biology and materials science for studying molecular systems ranging from small chemical systems to large biological molecules and material assemblies. The simplest calculations can be performed by hand, but inevitably computers are required to perform molecular modelling of any reasonably sized system. The common feature of molecular modelling techniques is the atomistic level description of the molecular systems. This may include treating atoms as the smallest individual unit or explicitly modelling electrons of each atom.

Molecular mechanics is one aspect of molecular modelling, as it refers to the use of classical mechanics/Newtonian mechanics to describe the physical basis behind the models. Molecular models typically describe atoms as point charges with an associated mass. The interactions between neighbouring atoms are described by spring-like interactions and van der Waals forces. The Lennard-Jones potential is commonly used to describe van der Waals forces. The electrostatic interactions are computed based on Coulomb's law. Atoms are assigned coordinates in Cartesian space or in internal coordinates, and can also be assigned velocities in dynamical simulations. The atomic velocities are related to the temperature of the system, a macroscopic quantity. The collective mathematical expression is known as a potential function and is related to the system internal energy (U), a thermodynamic quantity equal to the sum of potential and kinetic energies. Methods which minimize the potential energy are known as energy minimization techniques, while methods that model the behaviour of the system with propagation of time are known as molecular dynamics. Molecular modelling methods are now routinely used to investigate the structure, dynamics, surface properties and thermodynamics of inorganic, biological and polymeric systems. The types of biological activity that have been investigated using molecular modelling include protein folding, enzyme catalysis, protein stability, conformational changes associated with biomolecular function, and molecular recognition of proteins, DNA, and membrane complexes.

The book covers basic aspects of the subject, provides an example of a student research report. This book provides a separate chapter for each aspects of the subject.

— *Josef Carter*

Chapter 1

Magnetic Field

Intramolecular Forces

The electrical force which is responsible for all attractive interactions between atoms, molecules, and ions. The electrical force is described quantitatively by Coulomb's Law. A system in which forces act can possess two types of energy. These are kinetic energy, KE, and potential energy, PE. The magnitude of the kinetic energy is a function only of temperature, as we have seen. Potential energy arises from the action of the Coulomb force.

We will discuss the various ways in which Coulomb forces operate in atomic/molecular systems, manifesting in both intramolecular and intermolecular contexts. Intramolecular forces include the forces between atoms in covalent molecules; the forces between oppositely charged ions in ionic compounds; and the forces between identical atoms in metals.

Intermolecular forces include the forces that operate between one atom or molecule and another, or between the ion of one substance and the molecules of another. Although there are differences among the various types of forces, they are all fundamentally similar: they give rise to potential energy between the interacting particles, described by the potential well.

We shall see that an interplay between the depth of this well and the thermal kinetic energies of particles is responsible for many of the physical properties of substances.

Types of Energy and the Coulomb Force

An atom or a collection of atoms can possess energy of only two types. Kinetic energy (abbreviated KE) is energy of motion. A body of mass, m, moving with velocity, v, has kinetic energy given by equation.

$$KE = mv^2/2$$

When mass is expressed in kg and velocity in m/s, KE has units of kg m^2/s^2, orJoules (J). One Joule is defined as 1 kg m^2/s^2. The kinetic energy of atoms and molecules depends only on temperature, through the simple expression in 6-1-2.

$$\text{Average KE} = 3kT/2$$

Here Average KE is the average kinetic energy of a single atom or molecule; k is Boltzmann's constant; and T is the Kelvin temperature. otential Energy (abbreviated PE) is energy of position. A rock poised on top of a hill, even though it is not moving, has the potential to move if given a smll push. It therefore has potential energy. This potential energy is converted to kinetic energy as therock rolls down the hill. Allenergy is either kinetic or potential. The total enery of any mechanical system is the sum of its kinetic and potential energies:

$$E = KE + PE$$

Potential energy is an abstract concept; it is difficult to understand how something can possess energy simply by virtue of its position. A weight rests on a platform at height habove the ground. The weight is connected by a rope, across a pulley, to another platform that carries a load. If we remove the weight from the platform, it will fall, and in the process will lift the loaded platform. But in lifing the load a distance h, the weight has done work (a form of energy) on the load. To do work, the weight must use energy. It possesses the necessary energy to do the work by virtue of its position at distance h above the earth. Thus the concept of potential energy presupposes a force of attraction or repulsion. The weight falls to earth because there is a gravitational force of attraction between the earth and the weight. To move the weight from the ground to the platform initially, we would have had to use an amount of energy sufficient to overcome this force of attraction.

The amount of energy used to lift the weight is stored in the earth-weight system as potential energy. We retrieve this energy by allowing the earth and weight to come together. Potential energy is a consequence of a force. Our example has been based on the gravitational force because this is a familiar situation.

There are 3 forces other than gravitation. One of these is the electromagnetic force. This is the force by which electrons are attracted and held by the nucleus in an atom. The electromagnetic force is responsible for all interactions between atoms and molecules. It is far stronger than the gravitational force, which may be neglected in discussing atomic and molecular interactions (the electromagnetic force of repulsion between two electrons is 10^{42} times stronger than the gravitational force of attraction between them!).

The electromagnetic force operating at the atomic level is responsible for:

- The attraction of electrons to the atomic nucleus;
- The bonding together of atoms in a molecule;
- The attraction between oppositely charged ions in an ionic compound;
- The attraction between molecules that leads to liquefaction and solidification;
- The flow of electrical current;
- Your inability to push your finger or hand through a wall.

The magnitude of the electrical force is given by Coulomb's Law, equation.

$$F = q_1q_2/W\times r^2$$

q_1 and q_2 are the values of the two interacting electrical charges, in Coulombs, and r is the distance between them, in m. W is a constant that adjusts the units. Its value is 1.113×10^{-10} $C^2N^{-1}m^{-2}$. The force described by Coulomb's Law is responsible for all attractive forces between atoms, molecules, and ions. The potential energy associated with the Coulomb force is called the Coulomb potential energy.

Because force is the negative derivative of potential energy with respect to distance ($F = -dPE/dr$), the Coulomb potential energy can be obtained by integration of Coulomb's Law.

$$PE = q_1q_2/W\times r$$

When one of the charges is positive and the other negative, the potential energy is negative at all finite values of r.

Coulomb Forces and the Potential Well

Atoms, molecules, and ions exert attractive forces on one another because the atomic nuclei in one species attract the electrons in the other. These are called molecular forces, and the potential energy that results from the action of these forces is called molecular potential energy. We make several general statements about molecular forces and potential energy, and the relationship between them. The statements are equally valid for a pair of particles or a large collection of particles (here we are using he word, "particle", generically to represent an atom, a molecule, or an ion).

- Molecular forces are electrical;
- Molecular forces are attractive except at very close distances, and described by Coulomb's Law;
- The magnitude (strength) of the force between 2 particles varies with their distance of separation. The force is zero for an infinite separation, increases as the particles get closer, becomes a maximum when the particles just touch, and becomes strongly repulsive when the centers of the particles are separated by less than one particle diameter, due to interpenetration and repulsion of lectron clouds. The force has two components: one operates only at very small distance and is repulsive; the other acts at all distances and is attractive.

  Thus,

  $$F = q_1q_2/W\times r^2 + B$$

  B is a positive, repulsive force that operates only at very small distance.
- Action of the molecular forces draws the particles together (tends to ause them to aggregate) and results in a decrease in the potential energy of he pair (or collection) of particles. The force becomes stronger and the potential energy less until the particles touch.

  At distances less than 1 particle diameter, the force is repulsive and PE increases again. For all types of molecular forces, potential energy varies with distance of molecular separation, r. The plot is shaped like a valley, or well. It is usually called a *poential energy well.* The depth of the well is the enrgy that must be adde to increase the distance of separation between particles from r_1 to r_3 or greater.
- There are several types of molecular forces, which we will discuss shortly. All of them are characterized by a potential well of similar shape. However, the depth of the well increases with the strength of the forces.

Region: The particles, shown as two small circles just above the horizontal axis, are separated by large distances. Taking 2×10^{-10} m as a typical particle radius, the distances of separation in region 1 are 10 or more times larger. At large separation distance, molecular forces are nearly zero. For convenience, we take the potential energy to be zero when the forces are zero, so the PE curve coincides with the horizontal axis at large r. This situation of very large separation between atoms, molecules, or ions is a state of *deaggregation*. The gas phase of a pure substance is an example of a deaggregated state.

Region: The particles have approached closely enough to exert noticeable forces on one another. Action of the forces lowers the potential enrgy. The closer the particles approach (*i.e.*, the more aggregated they become), the stronger the forces and the lower the PE. The particles just touch. At this point, attractive forces are maximized and potential energy is at a minimum. This is an aggregated state. The solid phase of a pure substance is an example of an aggregated state.

Region: PE increases very rapidly as r decreases. In this region, the somewhat deformable electron clouds of the particles are being pushed into one another.

Since electrons repel one another, two particles that have interpenetrated exert a repulsive force on one another, and tend to fly apart. Region shows that when particles are forced together so that their centers are separated by less than one molecular diameter, repulsive forces, hence PE, rise very rapidly. This is the force that keeps you from falling through the floor.

The attractive forces are maximized and the PE minimized when the atoms, molecules, or ions just touch, at the bottom of the well. The distance corresponding to the potential energy minimum is the distance of closest approach. We have called this the aggregated state. In an ionic compound, such as KBr, this corresponds to a situation in which positive ions are surrounded by negative ions, and vice versa.

In a covalent compound, such as H_2O, the bottom of the well represents the situation in which the hydrogen atoms are covalently bonded to the oxygen atom, the bonds are separated by an angle of 104°, and the distance between the H and O atoms corresponds to r at the bottom of the well. A collection of He atoms or methane molecules can maximize the contactbetween themselves by aggregating such that a particular molecule touches as many other molecules as possible. This organization is referred to as close packing, and characterizes the solid state of a pure substance. PE is minimized in the aggregated state. Thus shows how potential energy changes as a substance is converted from the deaggregated to the aggregated state or vice versa. Notice that potential energy increases when an aggregated state is converted to a deaggregated one:

$$\text{Aggregated} \rightarrow \text{Deaggregated DPE} > 0$$

This is an idea that has broad applicability; we will encounter it many times. The phases (solid, liquid, and gas) of a pure substance fit the aggregation/deaggregation model very well. In the solid phase, the fundamental units of a substance are packed together in an orderly arrangement that minimizes potential energy. This is the aggregated state, in which intermolecular attractive forces are optimized. In the gas phase, the fundamental units are widely separated and experience only very weak forces from one another.

This is the deaggregated state. We might ask where the liquid state is on the diagram. From macroscopic observation, we know that the molar volume of a substace, *e.g.* water, in the liquid phase is very little different from the molar volume of its solid phase. These two aggregated (or condensed) phases have molar volume roughly 1000 times smaller than that of the gas phase. The liquid phase has distance of molecular separtion and potential energy only smewhat greater than those of the solid, at the point labelled r_2.

The condensed (aggregated) phases of matter and the conversions between the gas, liquid, and solid phases. Prior to that, it is important that you have an understanding of the types of forces that operate within and between molecules, and their relative strengths. This will enable you to understand, for example, why some substances exist as high-melting solids under normal conditions, while others exist as low melting solids, liquids, or gases.

Types of Molecular Forces

The nature and structure of a substance determine the type and strength of molecular forces that operate within it. We divide such forces into two broad categories. Intramolecular forces operate within the molecules or fundamental units of a substance. These are the covalent bonds within a molecule of a molecular substance, the forces between ions in an ionic compound, and the forces between atoms in a metal. Intermolecular forces operate between, rather than within, the molecules of a covalent substance; the atoms of a monatomic element; or the ions of one substance and the molecules of another. As a rule, intramolecular forces are much stronger than intermolecular forces. Both types of force, however, give rise to potential energy and a potential well. A listing of the important forces, arranged within these categories, follows:

Intramolecular Forces

- Attraction between positive and negative ions in a crystal of an ionic compound.
- Covalent bonds linking atoms in a network structure, such as that of silicon dioxide (sand).
- Covalent bonds in molecular substances.
- Metallic bonds

Intermolecular Forces

- Attractions exerted by one molecule of a molecular substance on another, such as the force of attraction between water molecules in ice.
- Attractions between atoms of the noble gas elements, helium through radon.
- Attractions between molecules of one substance and molecules of another, as when two liquids are mixed, or a molecular solid such as sugar is dissolved in a liquid.
- Attraction between molecules of one substance and ions of another, as when an ionic compound dissolves in a liquid.

In succeeding sections, we will discuss each of these molecular forces, and their relative magnitudes. We will begin with intramolecular forces.

Intramolecular Forces

The Ionic Bond

When two elements of widely different electronegativities combine, the result is an ionic compound. An example of an ionic compound is potassium bromide, KBr, in which the electronegativity difference between bromine, in group 17, and potassium, in group 1, is 2.14. Ionic bond formation involves transfer of an electron from the potassium atom, with its single valence electron and core charge, Z_{core}, of 1 unit, to he bromine atom, with its seven valence electrons and large Z_{core}

of 7 units. The electron transfer generates a full valence shell for each species. The ionic bond can be understood in terms of three energy erms. To remove an electron from the potassium atom to generate the potassium ion requires input of the first ionization energy of potassium.

The process is shown below:

$$K \rightarrow K^+ + e^- \quad I_1 = 420 \text{ kJ/mole}$$

I_1 for potassium is relatively small due to the minimum core charge found for group 1 elements. To add an electron to the bromine atom to form the bromide io releases an amount of energy equal to the first electron affinity of bromine.

$$Br + e^- \rightarrow Br^- \quad -A_1 = -325 \text{ kJ/mole}$$

The first electron affinity for bromine is substantial, because the large core charge of the bromine atom exerts a reasonable force on the extra electron. Despite this, to accomplish the electron transfer from potassium to bromine requires that 420–325 = 95 kJ/mole of energy be supplied. In other words, the electron transfer is substantially unfavourable! Thus we find that, contrary to a commonly held view of ionic bond formation, potassium does NOT "like" to transfer an electron to bromine; the process is unfavourable, and is not responsibl for the strength of the ionic bond. What, then, is?

Generation of positive and negative ions results in a strong attraction between them, according to Coulomb's Law. A single positive and negative ion are shown coming together to form an ion pair in reaction below.

$$K^+ + Br^- \rightarrow K^+Br^- \text{ (ion pair)}$$

Formation of one mole of ion pairs releases 589 kJ of energy, much more than was required to generate the ions. We conclude that it is the Coulomb attraction between oppositely charged ions that drives the formation of the ionic bond; this happens despite the need to form ions, which is unfavourable. In fact, ionic compounds do not consist of ion pairs of the type shown in reaction above. Rather, they consist of extended, highly organized aggregates in which positive and negative ions alternate. Figureshows both the ion pair and the lattice of potassium bromide.

The lattice is more stable than a collection of ion pairs because it surrounds each ion with ions of the opposite charge. This maximizes the attractive forces between positive and negative ions.

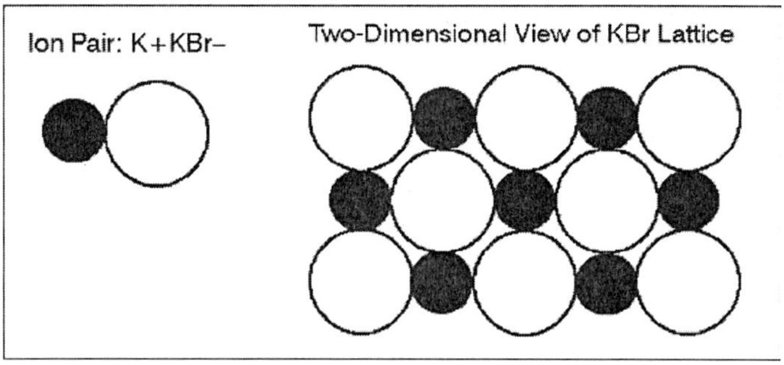

Figure: Potassium Bromide, Ion Pair and Lattice

In the ion pair, the positive charge of the potassium ion is offset on one side by the bromide ion, but is still exposed on all other sides. This exposed positive charge attracts bromide ions of other ion pairs, causing the ion pairs to condense to a lattice.

Formation of a lattice containing 1 mole each of potassium and bromide ions as in reaction releases 672 kJ of energy. This number, about 50 per cent larger than the amount released in formation of the equivalent amount of ion pairs, represents the depth of the potential well for potassium bromide.

$$K^+(g) + Br^-(g) \rightarrow KBr \text{ (lattice)}$$

The energy change that occurs when 1 mole of ionic substance in the solid lattice is formed from separated positive and negative ions in the gas phase is called the lattice energy, U, of the ionic compound.

The lattice energy is given by an expression having the form of:

$$U = KZ^+Z^-/r \text{ m}$$

Here K is a collection of physical constants, Z^+ is the cation charge, Z^- is the anion charge, and r is the distance of separation of the cation and anion in the lattice. Because the charge of the anion, Z^-, is negative, the lattice energy is a negative quantity, as expected.

This means simply that te solid lattice is more stable than the separated gas phase ions. Note that ths equation is essentially the expression for Coulomb potential energy. Without doing anycalculations at all, we can draw some important conclusions based on reaction.

- The potential well for an ionic compound of a 2+ cation with a 2- anion will be about 4 times deeper than that for a 1+/1- ionic cmpound, like KBr, due to the product of charges in the numerator of equation. Knowing the lattice energy for KBr allows us to estimate the lattice energy for MgO: about -2600 kJ/mole

- The potential well for an ionic compoun will be deeper when cation and anion are small than when they are large. Small cation and anionmeans small distance of separation, r, leading to a more negative U. The lattice energy for LiF is substantially more negative than that for KBr, despite the fact that both are 1+/1- compounds, ecause Li+ and F- are smaller ions. Lattice energies for a number of common ionic compounds are given in Table.

Table: Lattice Energies of Ionc Compounds

Charge Type	*Substance*	*U, kJ/mole*
1 + /1–	LiF	–1009
	NaCl	–774
	KBr	–672
	CsI	–584
2 +/2–	MgO	–2585
	BaO	–1978

- *Example:* The lattice energy for NaCl is -774 kJ/mole. The radii of the sodium and chloride ions are 116 pm and 167 pm respectively. Estimate the lattice energy for CsI, which has ion radii of 181 and 206 pm.
- *Solution:* NaCl and CsI are both 1+/1- ionic compounds. They should differ in lattice energy only through the r term of 6-4-5. Because the distance of separation is lager in CsI, the lattice energy should be less negative. We assume that r can be approximated by adding the radii of the ions. Thus for NaCl, r = 283 pm; for CsI, r = 387 pm. The lattice energies should be inversely related to the distance of separation of the ions:
  U(CsI)/U(NaCl) = r(NaCl)/r(CsI)
  U(CsI)/(–774) = 283/387
  U(CsI) = –570 kJ/mole

This compares reasonably well with the value in Table.

Example. Arrange the following ionic compounds in order of increasing intramolecular forces: KI, LiCl, MgO, CaO Al_2O_3.

Solution. KI and LiCl are 1+/1- ionic compounds. MgO and CaO are 2+/2- ionic compounds. Finally, Al_2O_3 is a 3+/2- compound.

Based on ion charges and equation, we can arrange these in three categories:

- KI, LiCl
- MgO, CaO
- Al_2O_3

The 1+/1- compounds have the least negative lattice energies. That for Al_2O_3 should be most negative, because the ion charges are largest. Within each category, the compound with the largest ions has the least negative lattice energy.

This reasoning leads to the following order:

$$U(KI) > U(LiCl) > U(CaO) > U(MgO) > U(Al_2O_3)$$

Here the greater than sign (>) is applied in the algebraic sense, in which numbers are considered to become smaller as they get more negative. In this convention, -50 is greater than–100. There are very few compounds that are legitimately considered to be ionic; that is, in which electron transferfrom metal to non-metal is complete. They are limited to compounds of the group 1 metals (Li, Na, K, Rb, Cs) with fluorine, chlorine, and oxygen. Compounds of the group 2 metals with oxygen, sulfur, and the halogens for typical ionic lattices, and are usually considered to be ionic compounds. However, there is considerable covalent character in the bonding.

Based on our discussion of the ionic bond in this section, it is understandable why ionic compounds are invariably solids at room temperature, with high melting and boiling points. The strong, non-directional Coulomb attractions acting between oppositely charged ions in the lattice can be overcome only at very high temperature.

The Covalent Bond

Two atoms of relatively large and similar electronegatvities (*i.e.*, non-metals) tend to form covalent compounds rather than ionic compounds. In covalent compounds, atoms share rather than transfer electrons to achieve full valence shells, because both atoms have similar substantial values of the core charge. Large Z_{core} tends to make electron transfer unfavourable. The shared electrons are paired, and are usually considered to be localized between the bonded atoms. In this section, we will discuss the chemist's view of the covalent bond. Two separated chlorine atoms, each with 7 valence electrons distributed in the 3s and 3p valence orbitals.

Each Cl atom has single odd electron in a 3p orbital. It is these electrons that will be pared and shared in the covalent bond. For the two atoms to bond by sharing electrons, it is necessary that they establish a mutual region in which the shared pair resides.

They accomplish this by merging, or overlapping, the atomic orbitals containing the electrons to be paired. Because these atomic orbitals have distinct spatial directions, the bonded atoms must maintain a definite orientatin relative to one another in order to preserve the overlap of orbitals. Thus covalent bonds are directional. The operation of Coulomb's Law in the covalent bond is perhaps less obvious than in the ionic bond. The two chlorine nuclei and the two electrons of the covalent bond, and indicates the various Coulomb forces that operate in this collection of centers.

There are several interactions: attrction of nucleus 1 for electron 1; attraction of nucleus 2 for electron 1; attraction of nucleus 1 for electron 2; attraction of nucleus 2 for electron 2; repulsion of nuclei 1 and 2; and repulsion of electrons 1 and 2. Note, however, that because the bonding electrns are required to be between the nuclei so that they can occupy the overlapping atomic orbitals, the nucleus-electron attractive forces are maximized, and the nuclear-nuclear repulsions are minimized, because the electrons shield the two nuclei from each other. Thus the strength of the covalent bond is attributable to the simultaneous attraction of both electrons to both nuclei.

The lengths and strengths of covalent bonds—the position and depth of the covalent bond potential well—vary with the bonded atoms. In general, thestrengths are substantial and comparable with those of ionic bonds. Thus to break all of the covalent bonds in a molecule of methane, CH_4, requires the input of 1662 kJ of energy. Methane lies in a deep potential well relative to the separated atoms. Generally speaking, the following generalizations hold for covalent bond strengths.

- Small atoms form relatively short bonds, and large atoms form relatively long bonds.
- Short bonds tend to be stronger than bonds formed by larger atoms because the bonding electrons are closer to the bonded nuclei. For example

Bond	*Length*	*Strength*
Cl–Cl	198 pm	240
Br–Br	228 pm	190
I–I	266 pm	149

- The word "tend" is emphasized here, because there are exceptions. Notably, single bonds of the small atoms, N, O, and F are very weak.
- Multiple bonds are stronger than single bonds. For example, the single, double, and triple bonds between carbon and oxygen have the following strengths:
  - C–O 358 kJ/mole
  - C = O 799 kJ/mole
  - C(3)O 1071 kJ/mole

This seems reasonable, because there are more electrons localized between the nuclei and attracted by both. Strengths and lengths of some typical covalent bonds are presented in Table. A more complete set of covalent bond strengths is found in Table below.

Table: Strengths and Lengths of Covalent Bonds

Bond	*Bond Energy (Strength), kJ/mole*	*Bond Length, pm*
Single Bonds		
H – H	432	74.2
H – F	565	91.8
H – Cl	428	127.4
H – C	411	109
H – N	386	101
H – O	459	96
C – C	346	154
C – F	485	135
C – Cl	327	177
C – O	358	143
C – N	305	147
N – N	167	145
N – O	201	140
N – F	283	136
O – O	142	148
O – F	190	142
Multiple Bonds		
C = C	602	134
C = O	799	120
C = N	615	
N = N	418	125
N = O	607	121
O = O	494	120.7
S = S	425	188.7
C(3)C	835	120
C(3)O	1072	113
C(3)N	887	116
N(3)N	942	109.8
P(3)P	481	189
As(3)As	380	

- *Example*: Arrange the following single bonds in order of increasing length. Then arrange them in order of increasing strength.

  H–O, O–Cl, Cl–Cl, H–Cl

- *Solution:* Longer bonds are formed by larger atoms; atoms become larger as we descend the periodic table. The sizes of the atoms involved here are ordered as follows:

  Cl > O > H

The bond lengths are then expected to be in this order: H-O < H-Cl < O-Cl < Cl-Cl. The strengths of bonds tend to decrease with increasing length.

Thus:

Cl-Cl < O-Cl < H-Cl < H-O.

Although simplified, this section gives us an appreciation for the origin and strengths of covalent bonds. We now turn to a problem with our view of the covalent bond having to do with molecular stereochemistry. *The prediction and rationalization of molecular stereochemistry (shape)*: The VSEPR theory. Thus VSEPR predicts the tetrahedral geometry for methane, in exact agreement with experiment. The interpretation of the covalent bond in terms of overlap of atomic orbitals of the bonded atoms requires that we understand molecular shape in terms of the distribution in space of the atomic orbitals of the central atom, which are presumably overlapped with suitable terminal atom orbitals in covalent bonds.

We pose this question: are the observed shapes of molecules consistent with the distribution in space of the atomic orbitals of the bonded atoms? If we ask this question of methane, the answer must be NO. On one hand, the C-H bonds are known to make 109.5 ° angles with one another. On the other hand, the s and p valence orbitals of carbon have spatial orientations that are not consistent with this angle. The s orbital is spherical, hence non-directional; and the p orbitals make mutual angles of 90°. In fact, the bond angles in most molecules in which 3, 4, or 5 electron groups surround the central atom are inconsistent with this 90° orbital orientation. This presents a problem for our view of the covalent bond. This problem was recognized in the 1920s, when the modern quantum theory was applied to the hydrogen atom and the spatial shapes of the atomic orbitals were discovered. Tetrahedral stereochemistry at carbon in methane had been proposed in 1874 by. Vant Hoff on the basis of the number of isomers produced when hydrogen atoms were replaced by different atoms. At that time, though, people knew nothing of atomic orbitals. Thus the new quantum theory was faced almost immediately with a challenge: explain the shape of methane. This problem was tackled and resolved by Linus Pauling, who made during his life almost innumerable substantial contributions t science.

Pauling proposed that the carbon atom uses hybrid orbitals to form four identical bonds to hydrogen atoms, directed to the vertices of a tetrahedron. His theory of hybrid atomic orbitals is still widely used today to rationalize (not predict) molecular stereochemistry. Applied to methane, the theory goes as follows.

In its most stable electron configuration, the carbon atom. The s orbital is fully occupied with an electron pair, and two of the three p orbitals are singly occupied. Based on this arrangement, we might expect carbon to form two bonds using the two odd electrons. However, this would fail to give it the octet. Thus we imagine that carbon moves, or "promotes", one of the 2s electrons to the empty p orbital. This prepares it to form 4 bonds.

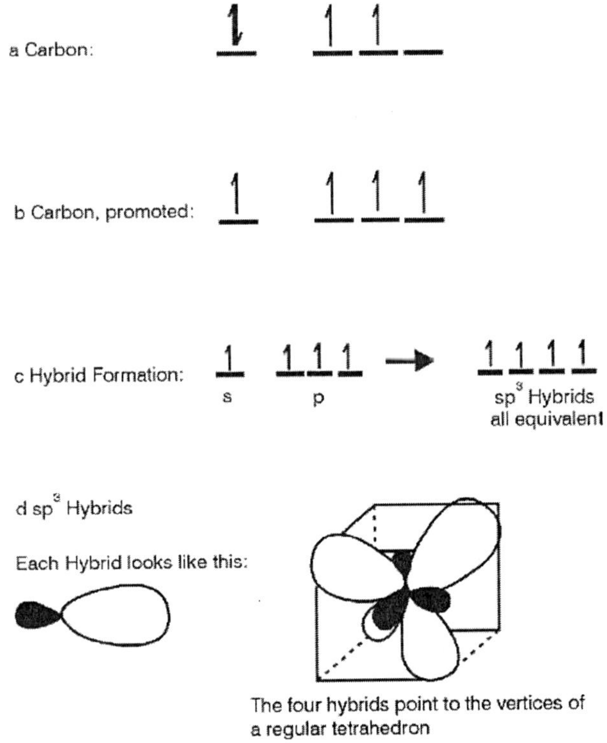

Figure: The Electron Configuration of Carbon

However, bond formation obviously does not occur using the pure atomic orbitals, because these can not produce a tetrahedral stereochemistry when overlapped with 1s orbitals of hydrogen atoms. Pauling proposed that the carbon atom creates from its 4 valence orbitals four new, equivalent orbitals called "hybrids." *The word hybrid means the same thing here as in agriculture*: A blend or combination of two strains of a species. He called these sp^3 hybrids to indicate that they are composed of a mixture of a single s and three p orbitals. Hybrid formation is shown schematically in figure.

Indeed, he showed mathematically that it is possible to combine the four s and p atomic orbitals to produce four new orbitals, and that these new orbitals are directed to the corners of a tetrahedron. The four hybrids are shown in Figure above. Pauling thus "squared" fact with quantum theory by proposing hybrid orbitals. Let's apply these ideas to a few more situations.

The phosgene ($COCl_2$) molecule is pictured in Figure. It has been experimentally determined that the molecule is trigonal planar (as expected from VSEPR), with a double bond between carbon and oxygen. How is the trigonal planar stereochemistry, involving 120° bond angles, to be rationalized with the 90° spatial distribution of carbon 2p orbitals?

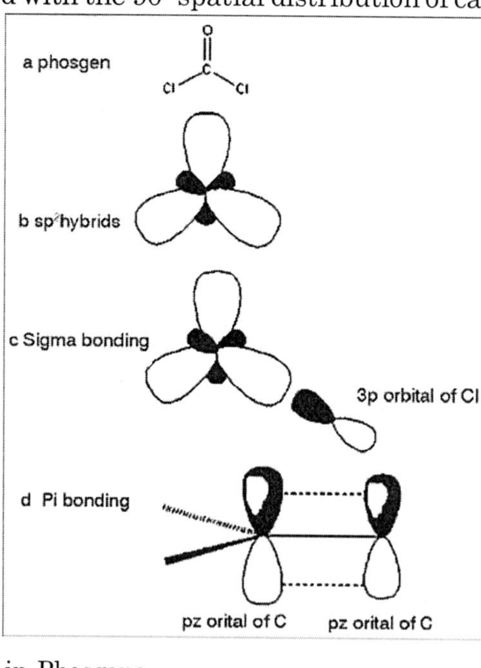

Figure: Bonding in Phosgene

Pauling showed that the 120° bond angle results if one assumes that C hybridizes the s and two of the three p orbitals to produce sp^2 hybrids.

Mathematically, the mixing of an s and two p's does indeed produce three equivalent new orbitals, directed to the corners of an equilateral triangle.

The plane in which the 3 hybrids lie depends upon which two p orbitals are used in the process. By convention, chemists normally choose to use px and py in the hybrids, which requires that the hybrids lie in the x-y plane of a Cartesian coordinate system. The pz orbital, not used in the hybrids, retains its usual shape and orientation along the z axis. Note, however, that it still contains an odd electron.

Once the hybrids are formed, each containing a single electron, carbon forms covalent bonds to the two hydrogen atoms and the oxygen atom. Each bond is formed by overlap of one carbon hybrid with one atomic orbital of the bonded atom (H or O). In forming these three bonds, orbitals are overlapped aong the line joining the bonded nuclei. Bonds involving overlap of this type are called *siga bonds*. At this point, bonding is not complete because carbon shares in only 7 electrons, as does oxygen. Each can complete the octet by forming one more bond, using the p orbitals oriented along the z axis.

Because these do not point at one another, but instead are parallel, overlap must occur in sideways fashion. Bonds formed by sideways overlap similar to this are called *pi bonds*. At this point, bonding is complete. We say that in phosgene, carbon uses sp^2 hybrids to sigma bond to H and O, and in addition forms a pi bond to O. The sp^2 hybrids are consistent with the trigonal planar sterochemistry. We are ready to make a few useful generalizations about hybrid orbital theory. Having done that, we will look at an example.

- The hybridization of an atom cannot be determined unless its stereochemistry is known. Hybrid orbitals cannot be used as a predictor of stereochemistry.

  CH_4 is tetrahedral at carbon; therefore, carbon uses sp^3 hybrids. NOT carbon uses sp^3 hybrids; therefore CH_4 is tetrahedral at C.

- The number of hybrids obtained is equal to the number of atomic orbitals used to make them. Thus, for example, there must be three sp^2 hybrids because three atomic orbitals (one s and two p) are used.

- An atom uses hybrid orbitals for its sigma bonds and lone pairs. Thus the number of hybrids required is equal to the sum of the number of lone pairs and sigma bonds:
- 4 sigma bonds + 0 lone pairs requires 4 hybrids, thus sp^3 (1 + 3 = 4)
- 3 sigma bonds + 1 lone pair requires 4 hybrids, thus sp^3 (1 + 3 = 4)
- 2 sigma bonds + 2 lone pairs requires 4 hybrids, thus sp^3 (1 + 3 = 4)
- 3 sigma bonds + 0 lone pairs requires 3 hybrids, thus sp^2 (1 + 2 = 3)
- 2 sigma bonds + 1 lone pair requires 3 hybrids, thus sp^2 (1 + 2 = 3)
- 2 sigma bonds + 0 lone pairs requires 2 hybrids, thus sp (1 + 1 = 2)
- Unhybridized p orbitals are used for pi bonds. The number of pi bonds equals the number of unhybridized p orbitals.

The flow of logic in determining hybrids used by the central atom is as follows. First, count the number of electron groups around the central atom; then apply VSEPR to determine the stereochemistry at that atom. The number of hybrids is the same as the number of electron groups counted for VSEPR.

A number of atomic orbitals equal to the required number of hybrids is used in forming the hybrids. The sum of the exponents in the hybrid designation indicates the number of hybrids of that type, and the number of atomic orbitals of each type used to make them.

Example: For each molecule, determine the Lewis structure; use VSEPR to determine molecular shape; and state the hybrid orbitals used by the central atom in bonding to the terminal atoms: $SiCl_4$; NH_3; BF_3; CO_2.

Solution: Lewis structures, then

Molecule	*#groups (hybrids)*	*Stereochemistry*	*Hybrids*
$SiCl_4$	4	Tetrahedral	sp^3
NH_3	4	Trigonal pyramidal	sp^3
BF_3	3	Trigonal planar	sp^2
CO_2	2	Linear	sp

Note that the number of pi bonds in BF_3 and CO_2 is the same as the number of unhybridized p orbitals on the central atom. It is relatively simple to extend the hybrid orbital concept to rationalize the stereochemistries of molecules in which the central atom has an expanded valence shell (more than 8 electrons). Essentially, this is done by incorporating in the hybrids as many valence d orbitals as required to give the necessary number of hybrid orbitals.

Example: Determine the hybrid orbitals used by the central atom in each molecule: PF_5, ClF_3, $XeOF_4$.

Solution:

Molecule	*# Electron Groups*	*Shape*
PF_5	5	Trigonal bipyramidal
ClF_3	5	T
$XeOF_4$	6	Square Pyramidal

Based on the numbers of groups, we deduce the hybrids. Five groups require 5 hybrids, which in turn require 5 atomic orbitals: one s, three p, and one d. Thus sp^3d. Six groups require 6 hybrids, which require the input of 6 atomic orbitals: one s, three p, and two d. Thus sp^3d^2. It can be mathematically demonstrated that 5 sp^3d hybrids point to the vertices of a trigonal bipyramid; and that 6 sp^3d^2 hybrids point to the vertices of an octahedron.

Before we leave the covalent bond and the hybrid orbital idea, something must be made clear. What is experimentally known is that carbon and hydrogen react to form methane, and that the methane molecule involves four carbon-hydrogen bonds that are perfectly tetrahedrally distributed in space. We introduced the ideas of electron promotion and orbital hybridization above as a way of rationalizing for ourselves, in terms of the output of quantum theory, what is experimentally observed. Thus promotion and hybridization are mental constructs. It is important that you realise that the carbon atom does not actually do these things. It is simply convenient for us to think of the carbon atom as doing these things.

There are numerous elements and compounds that consist of huge numbers of atoms covalently bonded together to give a large, 2- or 3-dimensional network. These substances are therefore similar to ionic compounds in that they do not consist of small, discrete molecules; however, they do not consist of discrete ions either. Examples of such substances are silicon dioxide (SiO_2), arsenic trioxide (As_2O_3), carbon in the diamond form, and phosphorus in the red form.

The large numbers of covalent bonds in these structures leads to potential well depths comparable to those found in ionic compounds.

Most such materials exist as solids at ordinary temperature, and melt and boil only at very elevated temperature.

Ionic and Covalent Bonding: A Continuum

The bond between two identical non-metal atoms, as in Cl_2, is purely covalent; that is, each atom has an exactly equal share in the bonding electron pair. Pure covalent bonds exist in the elemental forms of many of the non-metals (for example, all of the diatomic halogens of group 7; O_2 and S_8 of group 6; N_2 and P_4 of group 5; C and Si of group 4; and others). Bonds between different non-metal atoms are covalent, in that they involve electron sharing, but polar, meaning that they have some ionic character. The bonds between chlorine and the other elements of period 3 are shown below. The transition from pure covalent to polar covalent to ionic as we move from right to left in the period is clear.

Na-Cl—Mg-Cl—Al-Cl—Si-Cl—P-Cl—S-Cl—Cl-Cl

ionic—ionic—ionic—polar cov—polar cov—polar cov—cov

Although the Na-Cl and Mg-Cl bonds are usually considered to be ionic, they are not purely ionic; electron transfer from the sodium or magnesium atom to the chlorine atom is not complete. There is no such thing as a pure ionic bond. The process of electron transfer from a metal atom to a non-metal atom generates a positive ion and a negative ion. The positive ion is electron deficient, and attracts an electron pair of the negative ion strongly. Thus the very act of electron transfer creates a disparity that must then be offset by partial electron sharing. If the anion is one of the very electronegative non-metals (flurine or oxygen), it can resist the pull of the cation effectively, and hold on to the electron pair.

Anions other than F^- and O^{2-}, however, cannot completely resist the attraction of the cation for their lone pairs. These atoms share a lone pair with the cation to form a bond that, although it is very polar, is nonetheless partially covalent. As a general rule, the higher the charges on the cation and anion, the more covalency in the bond. A highly charged cation (3^+ or 4^+) is small and compact, and exerts a powerful attractive force on electrons. A highly charged anion (2^-, 3^-) holds its extra electrons weakly and readily shares them to some extent with the cation. The following series of compounds illustrates nicely the evolution from largely ionic to covalent bonding as cation charge increases:

Compound	*Physical State, 298 K*	*Melting Point °C*	
NaCl	Solid	801	
$MgCl_2$	Solid	714	
$AlCl_3$	Solid	190	
$TiCl_4$	liquid	-25	

The decrease in melting point indicates increasing covalency in the bonding. Titanium tetrachloride, $TiCl_4$, is a liquid at room temperature, indicating that it is molecular, not ionic. The Metallic Bond. Elements of similar but relatively low electronegativities (*i.e.*, metals) form metallic bonds. It is these bonds that are operative in a typical metal and that are responsible for metallic properties: reflectivity, conductivity, malleability, and strength. Metallic bonds form between atoms that have fewer valence electrons than they have valence orbitals. We shall try to develop a picture of the metallic bond by considering sodium metal, which exists as a very low melting, highly reactive solid at room temperature. Figure shows the Lewis dot symbols for two sodium atoms.

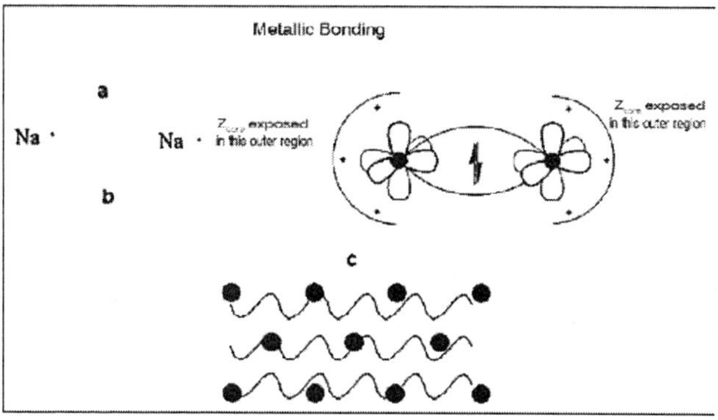

Figure: Metallic Bonding

The single odd electron is in the 3s valence atomic orbital of the sodium atom. At first glance, it might be expected that two sodium atoms, each with a single unpaired valence electron, would form a single covalent bond by overlapping the 3s valence orbitals and pairing the odd electrons in the resulting common region of space. There are problems with this bonding situation however.

First, bonding of this type would give each sodium atom only a duet of electrons, nowhere near the required octet. Each atom would have three empty 3p orbitals protruding from it into space. Second, pairing the 3s valence electrons localizes them in the region between the two nuclei.

They no longer have probability of being found in the peripheral regions opposite the bond. Because these are the only valence electrons, the regions marked with + symbols in the figure are exposed to the full core charge (+1) of the sodium atom. (The + signs in the figure do not represent actual positive charges; instead, they are intended to indicate that another atom located in this region would experience the core charge of the sodium atom near it). Here is the key point: another sodium atom in the region of the bonded pair would be drawn, via its valence electron, to the exposed core charge of the 2-atom cluster. It would contribute its valence electron to a bond with one or the other of the pair.

The sodium atoms are therefore driven to aggregate, using their relatively few valence electrons to bond together as many nuclei as possible via the Coulomb attraction. The aggregation is very regular in nature, resulting in an ordered arrangement of metal atoms called a *lattice*. As a result of aggregation, each sodium atom uses all of its valence orbitals in forming partial bonds with a number of neighbours.

Contrast this with the covalent bonding between two chlorine atoms. The covalent bond is again localized, but the remaining valence electrons are distributed evenly around each chlorine atom, effectively shielding the core charge from chlorine atoms in other molecules. Further, all of the s and p valence orbitals are full. In order to aggregate as sodium does, chlorine would have to use either the 3d orbitals or the orbitals from the $n = 4$ main shell. But the core charge experienced by an electron in either of these orbital sets is essentially zero ($17-17 = 0$). Consequently, chlorine is limited in its aggregation to the formation of covalently bonded diatomic molecules. As we will discuss shortly, there is only a very weak tendency for the diatomic molecules to aggregate via intermolecular forces. Interestingly, although hydrogen, like sodium, has only a single valence electron, it is not metallic. Instead, hydrogen atoms form covalently bonded diatomic molecules.

Why does hydrogen behave so differently from sodium and the other metals? Formation of a localized covalent bond between two hydrogen atoms results in exposure of the peripheral regions of the molecule to the core charge of the hydrogen atoms. In this case, because there are no core electrons, the H atom core charge is the nuclear charge of 1+. In the case of sodium and other metals, this situation leads to aggregation using the empty valence p orbitals.

However, hydrogen has no valence p orbitals. The single orbital in the n=1 shell is full; no further aggregation is possible. The formation of metallic bonds can occur only if empty valence orbitals are available. Experiments have shown that if hydrogen is subjected to extremely high pressure, which forces the diatomic molecules to aggregate, it exhibits metallic characteristics.

In sodium, as in most metals, there are too few electrons to form normal two-electron bonds between all pairs of bonded atoms. To achieve stability, the valence electrons are spread over many nuclei, simultaneously bonding them all together. This spreading of electrons is called *delocalization*, and in some ways is similar to the delocalization that occurs in resonance. In metals, however, delocalization is much more extensive, and it is best to think of the electrons as waves that extend over the entire dimensions of the metallic crystal. A schematic structure of metallic bonding.

The nuclei are represented as points, and the electrons as waves. This picture enables us to understand the properties of metals. That the electrons are delocalized waves is consistent with the conductivity of metals, and with their flexibility and strength. Putting stress on metals moves the nuclei, but the waves adjust, maintaining bonding. Because the electrons are waves, they are everywhere in the crystal at once. They are mobile, rationalizing the electrical and thermal conductivity of metals. Finally, the pooling of valence orbitals by aggregated sodium atoms leads to a large number of closely spaced orbitals in the metallic crystal. Thus the electrons absorb and reemit photons of all energies in the visible region of the spectrum, giving metals their highly reflective appearance.

There are so few bonding electrons in metallic sodium that only about 1/8 of an electron pair bond holds each pair of adjacent sodium atoms together. It is little surprise, then, that sodium is very soft (weak bonding), very reactive (deficiency of bonding electrons), with low density, and low melting and boiling points.

Example: In magnesium, how many metallic bonds form per Mg atom?

Solution: The electron configuration of the magnesium atom is $1s^2\ 2s^2\ 2p^6\ 3s^2$. We assume that metallic bonds are formed using the valence electrons only. Because each Mg atom contributes 2 valence electrons to the metallic bonding, and because 2 electrons constitute one bond in Lewis terms, there is one bond per Mg atom in the metallic lattice.

Intermolecular Forces

Having described the forces that operate within molecules (including ionic crystals and metals), we are ready to discuss the interactions that occur between species (atoms, molecules, and ions) that are not covalently bonded to each other. These are called *intermolecular forces*; they can originate in a number of ways.

Dipole-Dipole Forces

Intermolecular forces that operate between neutral molecules having molecular dipole moments are called dipole-dipole forces. The structure of a molecule may cause its centers of positive and negative charge to be displaced from one another, giving the molecule a dipole moment. The dipole moments of two neighbouring molecules tend to align with the + end of one dipole near the–end of the other, so that forces of attraction between them are maximized.

The maximum force of attraction between two dipoles, m_1 and m_2, separated by a distance r is given by equation.

$$F = \mu_1 \times \mu_2/\Omega \times r^4$$

Equation is Coulomb's Law, in disguised form. It is necessary to raise the distance of separation to the 4th power in the denominator to cancel the distance units that appear in both dipole moments in the numerator. Physically, the 4th power dependence means that the attraction between dipoles falls off much more rapidly with distance than would the attraction between isolated charges.

This is because the + and–ends of dipole 2 are almost equidistant from dipole 1 when the dipoles are separated by a distance that is much greater than the dipole length. The force of attraction between the + end of dipole 1 and the–end of dipole 2 is almost cancelled by the force of repulsion between the two + ends. Dipole-dipole forces operate between molecules of water, and between molecules of the substances.

Such forces are obviously much weaker than those operating in ionic or covalent network solids, and give rise to potential wells having depths in the approximate range 5-20 kJ/mole. Many molecular substances with dipolar molecules exist as liquids at ambient temperature, and have relatively low boiling points. In particular, many organic (carbo-containing) compounds are of this type.

Example: Which substances experience dipole-dipole intermolecular forces?

SiF_4, $CHCl_3$, CO_2, SO_2

Solution: We can draw the following conclusions about the molecules of these substances.

SiF_4	tetrahedral	Si-F bonds are polar, but no molecular dipole; bond dipoles cancel
CO_2	linear	C-O bonds are polar, but no molecular dipole; bond dipoles cancel
SO_2	bent	S-O bonds are polar and do not cancel. Sulfur lone pair dipole only partially offsets net bond dipole.
$CHCl_3$	tetrahedral	C-H and C-Cl bonds are polar and do not cancel

Hydrogen Bonding

Hydrogen bonding (H-bonding) is a special kind of dipole-dipole force that occurs when a hydrogen atom is bonded to one of the very electronegative atoms, F, O, or N.

The H-F, H-O, and H-N bonds are very polar, because the electronegative atom draws the bonding electron pair strongly to itself.

This leaves the hydrogen nucleus exposed. The resulting bond dipole is substantial. The positive end of this dipole, consisting essentially of an exposed hydrogen nucleus, has a high charge density because the proton is so small. Consequently, it strongly attracts the negative end of a similar bond dipole in another molecule to give a structure.

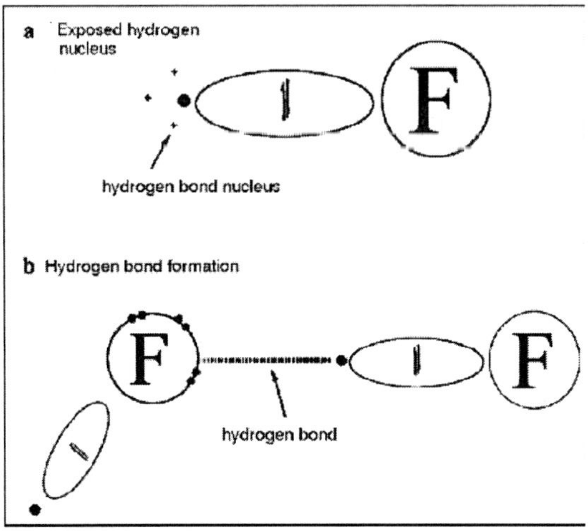

Figure: Hydrogen Bonding

The hydrogen atom is sandwiched between the two electronegative atoms in a linear arrangement. Hydrogen bond forces cause potential wells of depth in the range 5-50 kJ/mole.

Note that these are on average deeper than those resulting from ordinary dipole-dipole forces. The intermolecular forces between water molecules are of the hydrogen bonding type.

They are responsible for the abnormally high boiling point of water, which, with its small molecular weight, would otherwise be a gas at room temperature.

Compare the boiling point of water (100°C) with that of methane (-162°C), which has a similar molecular weight, but lacks hydrogen bonding and ordinary dipole-dipole forces. Hydrogen bonding is responsible for the formation of genes in the DNA molecule; for the helical structure of proteins; and for the incredible strength of Kevlar, a DuPont polymer used for canoe hulls and bullet-proof vests.

Example: In which substances do hydrogen bonding forces operate between molecules?

$$CHCl_3,\ CH_3CH_2OH,\ HNO_3,\ PH_3$$

Solution: Hydrogen is bonded to one of the very electronegative atoms in CH_3CH_2OH and HNO_3. Hydrogen bonding should occur in both of these substances.

Instantaneous Dipole-Induced Dipole (Dispersion) Forces

It may seem that there would be no forces of attraction between atoms of helium, because they are uncharged and have no permanent dipole moment. Yet He can be liquified at the very low temperature of 4K. That this is possible indicates that intermolecular forces do exist, though they are very weak. What is their origin? We portray the He atom as if the electron distribution about the nucleus is perfectly spherical. However, the electrons are constantly in motion and it is possible at any given instant for the electron cloud to be skewed slightly to one side of the nucleus.

This generates an instantaneous dipole moment in the He atom. During its transitory existence, this dipole induces a dipole in a neighbouring atom and the two instantaneous dipoles attract. Forces between transitory dipoles are called instantaneous dipole-induced dipole forces, or alternately, London dispersion forces after the scientist who first proposed them. We will refer to them as *dispersion forces.*

It is these forces that are responsible for the liquefaction and/or solidification of a number of substances whose molecules do not possess permanent dipole moments. For example, dry ice is solid CO_2. The CO_2 molecule is linear and, though it has non-zero bond dipoles, has no overall dipole moment.

Similarly, CCl_4, with perfectly tetrahedral molecules and no net dipole moment, is a liquid at room temperature and up to 78 °C. And iodine, which consist of I_2 molecules, is a solid at room temperature. Dispersion forces depend on two features of molecular structure. First, they increase in magnitude with the size and distortability (usually called the polarizability) of the electron clouds of the interacting particles. Size and polarizability increase as molecular weight increases.

It follows that dispersion forces increase as molar mass increases. For substances of large atomic or molecular mass, dispersion forces are strong enough that the substances are solid or liquid at room temperature.

We saw this above for CCl_4 and I_2. A striking example is provided by the diatomic halogens of group 17, which progress from F_2, a gas at room temperature and 1 atm pressure, to Cl_2, also a gas, to Br_2, a liquid, and finally to solid I_2. Only dispersion forces are operative in this series of substances. Second, dispersion forces depend upon molecular shape via the surface area over which two molecules can be in contact.

The larger the surface area of contact, the stronger the dispersion forces. Molecules that are roughly spherical in shape are able to contact each other only minimally. In contrast, molecules that are planar or linear in shape can maintain a large surface area of contact, with correspondingly larger dispersion forces. The classic example of the effect of molecular symmetry (shape) on the magnitude of dispersion forces is provided by the series of isomeric pentanes (C_5H_{12}).

Neopentane is highly symmetrical and nearly spherical in shape. Intermolecular forces are relatively small and the boiling point is correspondingly low.

Pentane is linear, so two pentane molecules are able to contact each other along the entirety of their length. Intermolecular forces are relatively large, and the boiling point is correspondingly high. Isopentane is intermediate structurally, and the least symmetrical of the three isomers. Intermolecular forces are less than in pentane, but greater than in neopentane.

The boiling point is intermediate. Graphite, the most stable form of elemental carbon at ordinary temperature and pressure, has the structure. Each carbon atom is bonded to three others in trigonal planar fashion, using sp^2 hybrid orbitals.

The unhybridized p orbitals are used to form a network of pi bonds that extends over the entire array of bonded carbons, producing a large planar sheet of carbon atoms. These sheets are then stacked one atop the other to produce the three dimensional structure of graphite. Interestingly, there are no covalent bonds holding one sheet to the next.

The sheets attract one another via dispersion forces, which are very strong because of the large surface area of contact. It is important to realise that dispersion forces operate between all molecules, whether or not other forces also operate. Molecules of chloroform, $CHCl_3$, are atttracted by a combination of dipole-dipole and dispersion forces. The magnitude of dispersion forces varies with molar mass; however, they are generally weaker than dipole-dipole forces, giving potential wells in the range 0.1 to 5 kJ/mole.

Example: Arrange the following non-polar molecules in order of increasing melting point.

$$SiF_4, CS_2, CI_4, GeCl_4$$

Solution: None of these molecules is expected to have a dipole moment. Only dispersion forces will operate, and these increase with increasing molecular mass. The molar masses of these substances follow:

Substance	*MM*
SiF_4	104.077
CS_2	76.131
CI_4	519.631
$GeCl_4$	214.402

The intermolecular forces, and the melting points, should increase in the following order:

$$CS_2 < SiF_4 < GeCl_4 < CI_4$$

The experimentally determined melting points are -110.8, -90, -49.5, and 171 °C, respectively.

Example: Molar masses, dipole moments, and boiling points for several covalently-bonded substances are given below. Rationalize the observed boiling points in terms of intermolecular forces.

Substance	*MM*	μ	*Boiling Point, K*
He	4.003	0	4
H_2	2.016	0	20
N_2	28.014	0	77
CO	28.010	0.33	83
HCl	36.461	3.60	188
HBr	80.912	2.67	206
SO_2	64.058	5.42	263
HF	20.006	6.37	292
CCl_4	153.823	0	350
H_2O	18.015	6.17	373

Solution: The substances fall into two categories, those with dipole moments and those without. Only dispersion forces operate in substances without dipole moments.

Selecting from the table those substances with no dipole moment, shown below, we see that, with the exception of hydrogen, boiling point increases with increasing MM, as expected from our discussion of dispersion forces above. The dipole moment of CO is so small that it seems reasonable to include it in this list. Its boiling point is only slightly larger than that of N_2, due to weak dipole-dipole forces in addition to the dispersion forces.

Substance	*MM*	μ	*Boiling Point, K*
He	4.003	0	4
H_2	2.016	0	20
N_2	28.014	0	77
CCl_4	153.823	0	350

Substances having substantial dipole moments are listed below, now in order of increasing dipole moment.

Substance	*MM*	μ	*Boiling Point, K*
HBr	80.912	2.67	206
HCl	36.461	3.60	188
SO_2	64.058	5.42	263
H_2O	18.015	6.17	373
HF	20.006	6.37	292

We see only a rough correlation of boiling point with dipole moment: HBr and HCl have boiling points on the low end of the scale; SO_2 is intermediate; and H_2O and HF have high-end boiling points. However, the boiling points of HBr and HCl are inverted based on the size of the dipole moment. HCl, with the larger dipole moment, has the lowest boiling point in the group of five. Here is clear evidence that dispersion forces operate in addition to dipole-dipole forces.

Dispersion forces are substantially larger for HBr (MM 80.9) than for HCl (MM 36.5). This is sufficient to raise the boiling point for HBr above that for HCl. The remaining 3 substances do not yield to the same argument, however. The MM of SO_2 is more than three times larger than those of HF and H_2O, and its dipole moment is only somewhat less than that for water.

Yet the boiling points of HF and particularly of H_2O are strikingly higher than that for SO_2. Here we have evidence of the potency of hydrogen bonding intermolecular forces, which operate in both HF and H_2O. Despite its small MM, HF boils 30° higher than SO_2 because hydrogen bonding causes the potential well of the liquid to be relatively deep. The boiling point for water is about 80° higher than that for HF because each water molecule can engage in up to four hydrogen bonds (one for each H atom, and one for each lone pair on the O atom).

Mixed Forces in Solutions

All of the forces that can operate at the molecular level within a single pure substance are discussed above. It is possible to have what we might call hydrid forces that operate when pure substances are mixed to form solutions. When an ionic solid such as sodium chloride dissolves in water, ion-dipole forces are responsible. The maximum magnitude of ion-dipole forces is given by a form of Coulomb's Law that is intermediate between equations.

$$F = q \times \mu/\Omega \times r^3$$

Here q is the charge on the ion, m is the dipole moment of the solvent, and r is the distance between the ion and the centre of the solvent dipole.

Example: Consider the following three situations:

- A charge of 1+ separated by 2.0×10^{-10} m from a charge of 1- (this might be two ions in a crystal);
- A charge of 1+ separated by 2.0×10^{-10} m from a dipole of magnitude 1.6×10^{-29} C-m (a dipole of this magnitude would result from charges of 1+ and 1- separated by a distance of 1.0×10^{-10} m).
- Two dipoles, each of magnitude 1.6×10^{-29} C-m, separated by 2.0×10^{-10} m.

Compare the magnitudes of the attractive forces operating in these situations. This will provide an idea of the relative magnitudes of ion-

ion, ion-dipole, and dipole-dipole forces. Then, compare the forces when the distance of separation of the interacting entities is increased by a factor of 4 to 8.0×10^{-10} m.

Solution: We apply equations:

- *Ion-ion*: $F = (1.6 \times 10^{-19}\ C)^2/W \times (2.0 \times 10^{-10})^2 = 6.4 \times 10^{-19}/W$ Newtons
- Ion-dipole, $F = (1.6 \times 10^{-19})(1.6 \times 10^{-19})/W \times (2.0 \times 10^{-10})^3 = 3.2 \times 10^{-19}/W$ Newtons
- Dipole-dipole $F = (1.6 \times 10^{-29})^2/W \times (2.0 \times 10^{-10})^4 = 1.6 \times 10^{-19}/W$ Newtons

The ion-ion force is stronger by a factor of 4 than the dipole-dipole force. The ion-dipole force is intermediate. When the separation distance is increased by a factor of 4, the magnitudes of all forces decrease because they depend inversely on a power of the distance. The ion-ion force is reduced by a factor of 4^2; the ion-dipole force by a factor of 4^3; and the dipole-dipole force by a factor of 4^4:

- *Ion-ion*: $F = 4 \times 10^{-20}/W$ NtIon-dipole $F = 0.5 \times 10^{-20}/W$ NtDipole-dipole $F = 0.063 \times 10^{-20}/W$ Nt

The forces involving dipoles clearly decrease much more rapidly with increasing distance of separation than do ion-ion forces. Whether or not two liquids are mutually soluble (miscible) depends on whether the mixed intermolecular forces can compete with the forces operating within each pure liquid alone. The intermolecular forces in pure water are primarily hydrogen bonding (dipole-dipole) forces, superimposed on weak dispersion forces.

Those in pure acetone are dipole-dipole forces superimposed on dispersion forces that are stonger than the dispersion forces in water. When acetone and water are mixed, they readily dissolve in one another in all proportions, apparently because the dipole-dipole forces between water and acetone molecules are of strengths comparable to those operating in the two separate liquids. On the other hand, only dispersion forces operate within pure carbon tetrachloride, because the CCl_4 molecule has a molecular dipole moment of zero. Because the molecules are spherical, giving small surface area of contact, these forces are fairly weak.

Consequently, carbon tetrachloride and water are not mutually soluble. They are said to be *immiscible*. When CCl_4 and water are mixed, the two liquids segregate into layers, with the denser liquid, chloroform, on the bottom. The liquid-liquid interface (boundary) is clearly visible. Finally, a non-polar solid like I_2 dissolves readily in a non-polar liquid like CCl_4, because the mixed dispersion forces are comparable in magnitude to the dispersion forces in pure CCl_4 and pure I_2. However, the solubility of I_2 in water is essentially zero, because mixed forces cannot compete with the strong dipole-dipole forces in water.

Water molecules are unwilling to separate in order to make room for the non-polar I_2 molecules. In this way, we can extend our intermolecular force ideas to provide an understanding of the formation of mixtures.

Mixed-Force Concepts

- Substance A will dissolve in substance B if the forces operating between A and B molecules are of similar type (dipolar or dispersion) and strength to those operating between A molecules and between B molecules. This statement is often sloganized: "Like dissolves like."
- Substance A will not dissolve in substance B if the forces operating between A and B molecules are of different types (one dipolar, one dispersion) (and therefore usually different strengths) than those operating between A molecules and between B molecules. This can also be sloganized: "Like does not dissolve unlike."

Example: For each pair, predict whether substance A will dissolve in liquid B, and explain your reasoning.

Substance A	*Liquid B*
C_2H_5OH	CH_3OH
$Br_2(l)$	C_5H_{12}
$Br_2(l)$	H_2O
C_2H_5OH	H_2O
C_2H_5OH	$CHCl_3$
NaCl	$CHCl_3$
NaCl	CH_3CH_2OH
H_2O	CH_2Cl_2

Solutionl: C_2H_5OH (ethanol) and CH_3OH (methanol) are both liquids due to fairly strong H-bonding intermolecular forces. Ethanol is expected to dissolve in methanol. Br_2(l) and C_5H_{12} (pentane) are both non-polar liquids with dispersion forces operating between molecules.

Thus they are "like" and we expect bromine to dissolve in pentane. Br_2 and H_2O are "unlike." Bromine will not dissolve readily in water.

Both liquids exert H-bonding forces at the molecular level. Ethanol is expected to dissolve in water.

This is not obvious. At first glance, we would call ethanol and chloroform "unlike" because H-bonding forces operate in ethanol, whereas dispersion forces operate in chloroform. However, in addition to its polar -OH group, ethanol has a non-polar hydrocarbon portion that is capable of exerting dispersion forces on molecules of chloroform. These two liquids are therefore miscible. Ionic NaCl will certainly not dissolve in non-polar chloroform.

Because ethanol is polar at the -OH end, we might expect NaCl to dissolve to some extent in ethanol, but not to the same extent as in water. This is indeed found. Polar water will not mix with non-polar dichloromethane.

Example: Describe what the system will look like during and following each process:

- 50 mL of distilled water is added to an Erlenmeyer flask.
- 0.10 g of I_2(s) is added to the flask, and the flask is stoppered and vigourously shaken.
- 50 mL of CCl_4 is added to the Erlenmeyer flask, and the flask is stoppered and shaken.

Solution:

- The clear, colourless liquid will occupy the bottom of the flask.
- I_2, and intensely violet solid, is not expected to be soluble in water because of incompatible intermolecular forces. Because I_2 is more dense than water, it will sink to the bottom of the flask. The water will remain clear and colourless.
- CCl_4 is a non-polar liquid in which the intermolecular forces are of the dispersion type. It is expected to be immiscible with water. Because it is more dense than water, it will sink to the bottom of the flask, and the water will float on it with a distinct liquid-liquid boundary. CCl_4 and I_2 have compatible intermolecular forces, so I_2 is expected to dissolve readily in carbon tetrachloride to give a purple solution. The flask will thus contain a layer of purple liquid under a layer of colourless liquid.

Distribution of a solute Between Two Immiscible Liquids in Contact

When two immiscible liquids, such as water and chloroform ($CHCl_3$), are placed in contact, they will mutually exclude each other, forming two liquid layers. The less dense liquid will float on the denser one, and a boundary surface will be clearly visible. It is interesting to think about what will happen if a third substance, solid or liquid, is added to the system consisting of two immiscible liquids, A and B.

If the third substance, C, is soluble in one of the liquids (say A) and insoluble in the other (B), it will dissolve exclusively in the A layer, as in example 6-14. More interesting is the situation in which the C is soluble in both liquids to at least some extent. In this case, we find that the substance C *partitions*, or *distributes*, itself between the two liquid layers. We further find that the concentration of the third substance is higher in the liquid with which its intermolecular forces are most compatible. The ratio of the concentration of C in liquid A to that in liquid B is found to attain a constant value K_d, defined as in equation:

$$K_d = [C]_A/[C]_B$$

This constant ratio is established no matter how the 3-component mixture is prepared.

Three such ways follow:

1. A and B are first brought into contact. C is then added, and the container is shaken.
2. C is first dissolved in A. A (containing C) and B are then brought into contact, and the container is shaken.
3. C is first dissolved in B. A and B (containing C) are then brought into contact, and the container is shaken.

In addition, the ratio is reestablished if the mixture is altered in any of the following ways:

- More A is added.
- More B is added.
- More C is added.

We can understand this behaviour in terms of a simple layers of A and B in contact at a boundary, and molecules of C initially dissolved in liquid A. The boundary between the two liquids provides a “doorway” from one to the other through which C molecules, which have reasonable interactions with both A and B molecules, can pass.

Thus molecules of C will tend to move across the boundary from liquid A, where they are initially dissolved, into liquid B, in which there are initially no molecules of C. In other words, they begin to distribute themselves between the two liquids. After some time, enough C molecules will have passed into liquid B that the rates of passage of C from A to B and back will be the same. Once this situation is attained, there will be no further change in the concentration of C in the two layers. In other words, the constant ratio in equation will be attained. From the magnitude of K_d, we may easily deduce which liquid C prefers. If $K_d < 1$, C prefers B; if > 1, C prefers A.

Distribution of a substance between two liquid phases is an example of dynamic equilibrium, a situation in which two opposing processes (here, movement of C both ways across the boundary) occur at equal rates.

A Nod to Other Classification Systems

Classification of the various types of forces as "inter-" or "intramolecular" is not absolute.

Thus you may see the forces classified as follows:

- Covalent Bonds (in molecular and network covalent substances)
- Ionic Bonds
- Metallic Bonds
- Interunit forces
  - Ion-dipole
  - Dipole-dipole
  - Hydrogen bonds
  - Dispersion forces

In this classification scheme, ionic and metallic bonds are categorized separately rather than as intermolecular forces, because there are no discreet molecules in ionic compounds or metals. Another classification system has recently become popular and is experiencing increasing use. This goes as follows.

- Covalent Interactions (includes covalent bonds in molecular and network covalent substances)
- Non-covalent Interactions
  - Ionic bonds
  - Metallic bonds

- Metal-ligand interactions
- ion-dipole forces
- dipole-dipole forces
- hydrogen bonds
- dispersion forces

Here there are only 2 major categories: covalent, and not covalent! It is important to realise that the distinction between covalent bonding and some of the non-covalent interactions is blurred.

For example, many people believe that metal-ligand interactions and even hydrogen bonding are weak covalent interactions.

The bottom line is that it is not particularly important how the classification is done, as long as the various types of interactions are understood. Pick your own favourite classification system, and use it!

The Interplay Between Molecular Kinetic and Potential Energies

The concept of temperature is of profound significancein science. We take it for granted in our daily lives, without thinking much about what it is. Temperature is the manifestation of molecular kinetic energy. The steam rising from a pan of boiling water feels hot because the molecules of water vapour are moving rapidly. Their collisions with the skin transfer kinetic energy, which we interpret as heat. The relationship between temperature and the average translational kinetic energy of an atomic, ionic, or molecular unit is very simple.

$$\text{Average } KE_{molecule} = 3/2\ kT$$

T is the absolute (Kelvin) temperature, and k is a fundamental constant of nature called Boltzmann's constant. Its value is 1.381×10^{-23} J/K. By $KE_{molecule}$, we signify the kinetic energy of any microscopic unit of matter.

This can be anything from an electron to a DNA molecule. Keep in mind in using equation that the kinetic energy referred to is the translational kinetic energy—the energy of motion of the particle as a whole through space. Translational kinetic energy is often called thermal energy.

Example: Calculate the thermal energy of one mole of nitrogen gas at room temperature, 293 K.

Solution: Calculate the kinetic energy of an average molecule; then multiply by Avogadro's Number.

Average $KE_{molecule}$

$= 3kT/2$

$= 3(1.381 \times 10^{-23}\ J/K)(293\ \ K)/2$

$= 6.07 \times 10^{-21}$ J/molecule

KE_{mole}

$= N_o KE_{molecule}$

$= 3N_o kT/2$

$= 6.07 \times 10^{-21}$ J/molecule$\times 6.023 \times 10^{23}$ molecules/mole

$= 3.66 \times 10^{3}$ J/mole

The product of Boltzmann's constant and N_o is a constant, R, which occurs in the ideal gas law. Its value is 8.314 J/K-mole. Expressed in terms of R, the kinetic energy per mole of gas is 3RT/2. We have seen that molecular forces promote aggregation by lowering molecular potential energy. Kinetic energy counteracts the effects of intermolecular forces. These ideas lead to a very useful generalization:

Potential energy favours aggregation; kinetic energy opposes aggregation. The stronger the molecular forces, the more rapidly atoms, molecules, or ions must move in order to overcome them. Even the strongest forces, however, may be overcome by the vigour of molecular motion if the temperature is high enough. At a temperature that gives a kinetic energy comparable in magnitude to the depth of the potential well, molecules begin to escape the potential well, and the substance undergoes a phase change (that is, sublimes, melts, or boils).

Table: Melting Temperature and Potential Well Depth

Substance	*Potential Well Depth, J/mole*	*Melting T, K*
He	0.105	3.5
H_2	1.04	14.0
Xe	14.7	161
N_2	6.30	63.2
Cl_2	26.8	172.1
I_2	56.3	387
H_2O	46.7	273
CO_2	33.6	217
NaCl	771	1074

A phase change may therefore by viewed as an exchange of KE for PE. If the potential well is shallow, as in liquid helium (where the only forces are the very weak dispersion type) the required kinetic energy is achieved at very low temperature (4 K). If the potential well is deep, as in the ionic compound MgO (where the forces are the strong ion-ion type), very high temperature is required for particles to escape

(3125 K). Table shows the parallel between the potential well depth and the temperature of melting for a number of substances.

Try plotting melting point against potential well depth to see how good the correlation is.

Non-Ideality of Gases

The postulates of KMT define an ideal gas. An ideal gas obeys the ideal gas law under all conditions of T and P. There is in fact no real gas that satisfies this definition. The ideal gas is a concept that we can think about but which does not actually exist.

Consider a sample of N_2 gas at 25 °C and 1 atm pressure. Imagine measuring the volume of the gas periodically as we cool it at constant pressure. If N_2 were ideal, it would follow Charles' Law exactly even to absolute zero. A plot of V versus T for ideal N_2 would follow curve A in Figure.

The actual V-T plot for N_2 follows curve A to a temperature of about 100 K. At this point the plot starts to curve down (curve B). At 77 K the volume sharply decreases while the temperature remains constant, and liquid appears in thecontainer. When all gas has converted to liquid, further decrease in T has little effect on the volume occupied by the liquid until T = 63 K. At this point, N_2 solidifies with a slight decrease in volume. Further cooling causes negligible change in V.

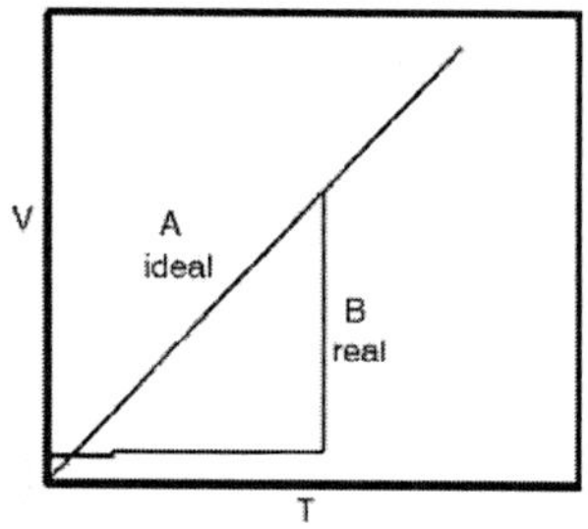

Because all real substances liquify at some T, no real substance can obey the ideal gas law under all P,T conditions. Deviation of real gases from ideal behaviour is shown conveniently in terms of the compressibility factor, Z

$$: Z = PV/nRT$$

Z is plotted against P in figure. For an ideal gas, Z = 1 at all pressures (the horizontal dashed line). Two deviations from ideality are commonly observed. In one type, Z always exceeds 1. H_2 is an example. In the second type, Z < 1 at moderate pressure but becomes

> 1 at high pressure. N_2 is an example. Most gases show behaviour similar to that of N_2. Real gases deviate from ideality because they violate two postulates of the KMT.

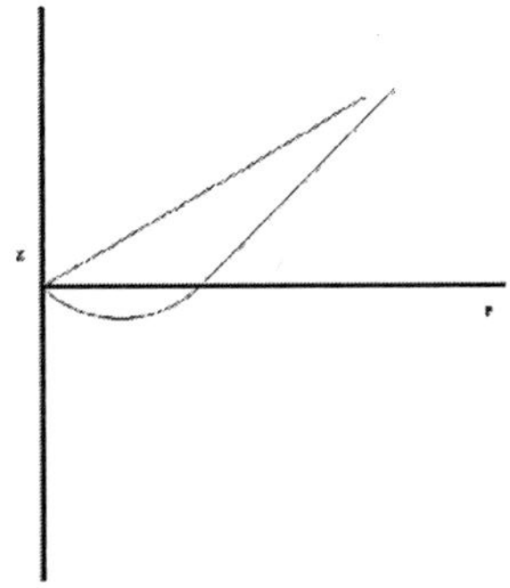

Figure: The Compressibility Factor, Z

First, real molecules exert attractive forces on one another. The forces cause the molecules to temporarily stick together in pairs when they collide, reducing the effective number of free particles in the gas. This is shown in Figure. Although there are 7 gas molecules in the container, 4 of them are temporarily stuck together in pairs, reducing the effective number of independent gas particles to 5. Sticking together reduces the pressure below the ideal value because there are fewer collisions with the container walls per unit time.

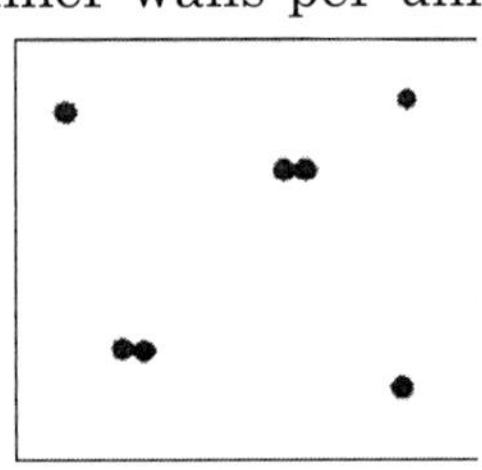

Figure: Non-Ideality of Gases

$$\text{Thus, } P_{act} = P_{ideal} - DP$$

Here P_{act} is the actual pressure of the gas, P_{ideal} is the pressure it would exert in the absence of intermolecular forces, and D P is the pressure reduction. If there are n moles of gas in a container of volume V, the pressure reduction, D P, is proportional to $(n/V)^2$.

The following argument demonstrates this. First, we assume that D P is proportional to the number of pairs of stuck-together molecules:D P proportional to (# pairs). How many pairs are possible for N molecules? If N = 2, there is only 1 pair. If N = 3, 3 pairs are possible. For N molecules, there are N(N-1)/2 possible pairs, where division by 2 avoids double counting. If N is very large, as in a macroscopic gas sample, N-1 = N and the number of pairs is $N^2/2$. Dividing by volume to convert to concentration and by N_o to convert molecules to moles gives,

D P proportional to (# pairs) proportional to,

$$(n/V)^2 = a(n/V)^2$$

where "a" is a proportionality constant. Substitution in equation gives,

$$P_{ideal} = P_{act} + a(n/V)^2$$

Second, real molecules are not points, but instead occupy finite volume. Thus a small fraction of available container space is occupied by the molecules themselves, and is not available to accommodate motion of the molecules. The total container volume is the sum of the volume of the molecules and the free volume available for motion:

$$V_{container} = V_{available} + V_{molecules}$$

The volume of the molecules is constant for a given gas, and depends on molecular size. As $V_{container}$ is made smaller, $V_{available}$ becomes a smaller fraction of $V_{container}$. Since the molecules have smaller free volume than they would if ideal (when $V_{molecules} = 0$), they exert a larger than ideal pressure. Equating $V_{container}$ with V_{actual} and $V_{available}$ with V_{ideal}, we obtain,

$$V_{act} = V_{ideal} + V_{molecules}$$

The volume of the molecules is proportional to the number of molecules, which is in turn proportional to the moles of gas: $V_{molecules}$ proportional to n = bn where b is a proportionality constant roughly equal to the volume per mole of substance. (The volume of the molecules can be determined by crowding the molecules together—*i.e.*, by liquifying the substance. The molar volume of the liquid phase gives an estimate of "b". The molar volume of liquid is the product of the MW and the reciprocal of density: b(vol/mole) = MM(mass/mole)×1/r (vol/mass) = MM/r (vol/mole) Using this approach, b should be 18/1.0 = 18 mL/mole for water.) Substituting equation rearranging for V_{ideal} gives

$$V_{ideal} = V_{act} - bn$$

Finally we substitute equations into the ideal gas equation, $P_{ideal}V_{ideal} = nRT$ to obtain equation:

$$(P + an^2/V^2)(V - nb) = nRT$$

This equation was first proposed by van der Waals in 1873 to explain deviations of gases from ideal behaviour. Its success earned him the Nobel Prize for Physics in 1910. In real gases the effects of non-zero molecular volume and intermolecular forces oppose. The former causes pressure to be higher than ideal, and the latter causes it to be lower than ideal. For N_2 and many other gases having $Z < 1$ at moderate pressure and $Z > 1$ at high pressure, first one effect, then the other, takes precedence. At low pressure (10-100 atm), the intermolecular forces dominate because the molecules are not yet compressed enough for the volume effect to be felt. This makes $P_{act} < P_{ideal}$, and $Z < 1$. As P becomes very large (> 100 atm), non-zero molecular volume becomes important, making $P_{act} > P_{ideal}$ and $Z > 1$. For hydrogen, intermolecular forces are small and the volume effect dominates at all pressures.

At low pressure (approximately 1 atm) and temperature much higher than the boiling point of the substance, both non-ideal effects are small and real gases approach ideal behaviour. Values of the van der Waals constants a and b for several gases are in Table.

Table: Van der Waals Constants for Gases

Gas	*a, L^2-atm/mole*	*b, L/mole*	
Ar	1.35	0.0322	
Cl_2	6.49	0.0562	
CO_2	3.59	0.0427	
H_2	0.244	0.0266	
He	0.034	0.0237	
N_2	1.39	0.0391	
O_2	1.36	0.0318	

Example: Discuss the Van der Waals constants in Table in terms of molecular structure.

Solution: The constant, a, is related to the magnitude of intermolecular forces; b is related to molecular volume. As a first guess, we might expect both of these to get larger with increasing molar mass of the gaseous substance.

Let us predict that the constants will parallel molar mass. The substances and their constants are arranged in order of increasing MM below.

In a rough sense, our prediction is borne out. Both a and b trend upward as molar mass increases. However, we are wrong in the details. The value of a for He is a clear exception; and the constants for nitrogen, oxygen, and argon do not parallel the molar masses. Having predicted, somewhat but not entirely successfully, we now must rationalize (*i.e.*, try to think of explanations for deviations from the trend).

Gas	*MM*	*a, L^2-atm/mole*	*b, L/mole*
H_2	2.016	0.244	0.0266
He	4.003	0.034	0.0237
N_2	28.014	1.39	0.0391
O_2	31.998	1.36	0.0318
Ar	39.948	1.35	0.0322
CO_2	44.009	3.59	0.0427
Cl_2	70.906	6.49	0.0562

First, we examine the hydrogen, helium pair. The difference between these is that hydrogen is molecular, and helium is atomic. Hydrogen molecules are highly symmetrical, but elongated from a perfect sphere. The hydrogen molecules are more susceptible to dispersion forces than are the spherical, compact (high core charge) helium atoms. Consequently, forces (and the constant a) are larger for H_2, despite its lower molar mass.

Incidentally, because there are 2 atoms per particle in hydrogen, the volume constant, b, is slightly larger for hydrogen. Next, we consider nitrogen, oxygen, and argon, which have roughly similar molar masses. Again, Ar is atomic and spherical, whereas nitrogen and oxygen are molecular, with elongated electron distributions. Despite their greater mass, the argon atoms are less susceptible to dispersion forces because they are spherically shaped. For the same reason, the volume constant for Ar is comparable to those for nitrogen and oxygen, despite the greater number of electrons. Atomic size decreases left to right in a period; oxygen atoms are actually smaller than nitrogen atoms, despite their larger atomic weight. Thus the volume constant, b, is smaller for oxygen than for nitrogen. Oxygen is more electronegative than nitrogen, making it somewhat less distortable. Intermolecular forces, measured by a, are somewhat less in O_2.

Electromagnetic Theory

James Clerk Maxwell, who had set himself the task of mathematically working out Michael Faraday's views, and who, both by doing so and by introducing many new ideas of his own, became the founder of the modern science of electricity,' recognized that, at every point of an electromagnetic field, the state of things can be defined by two vector quantities, the "electric force " E and the " magnetic force " H, the former of which is the force acting on unit of electricity and the latter that which acts on a magnetic pole of unit strength.

In a non-conductor (dielectric) the force E produces a state that may be described as a displacement of electricity from its position of equilibrium. This state is represented by a vector D ("dielectric displacement") whose magnitude is measured by the quantity of electricity reckoned per unit area which has traversed an element of surface perpendicular to D itself. Similarly, there is a vector quantity B (the "magnetic induction") intimately connected with the magnetic force H.

Changes of the dielectric displacement constitute an electric current measured by the rate of change of D, and represented in vector notation by C==D Periodic changes of D and B may be called "electric " and "magnetic vibrations." Properly choosing the units, the axes of coordinates (in the first proposition also the positive direction of *s* and n), and denoting components of vectors by suitable indices, we can express in the following way the fundamental propositions of the theory.

- Let *s* be a closed line, a a surface bounded by it, n the normal to a. Then, for all bodies, *H 8 ds =f C n dcr,* J *Esds =-d* f *Bnda,* where the constant *c* means the ratio between the electro-magnet and the electrostatic unit of electricity. From these equations we can deduce: For a surface of separation, the continuity of the tangential components of E and H. The solenoidal distribution of C and B, and in a dielectric that of D. A solenoidal distribution of a vector is one corresponding to that of the velocity in an incompressible fluid. It involves the continuity, at a surface, of the normal component of the vector.
- The relation between the electric force and the dielectric displacement is expressed by Dx = eiEx, Dy = e2Ey, D z = E3Ez, the constants (dielectric constants) depending on the properties of the body considered. In an isotropic medium they have a common value *e,* which is equal to unity for the free aether, so that for this medium D = E.

- There is a relation similar to between the magnetic force and the magnetic induction. For the aether, however, and for all ponderable bodies with which this article is concerned, we may write B =H.

It follows from these principles that, in an isotropic dialectric, transverse electric vibrations can be propagated with a velocity v = $c/\text{-}J\,E$. Indeed, all conditions are satisfied if we put Dx =o, Dy =*a* cos *n(t-xtr'+l), 'D z =o, H z =o, H y =0, Hz=* avc 1 *cos* n(t–xa+*l*) For the free aether the velocity has the value c.

Now it had been found that the ratio *c* between the two units of electricity agrees within the limits of experimental errors with the numerical value of the velocity of light in aether. (The mean result of the most exact determinations 2 of c is 3,001 to 10 cm./sec., the largest deviations being about 0.008 10 10 :. and Cornu 3 gives 3,001 to 10 0,003 to 10 as the most probable value of the velocity of light.) By this Maxwell was led to suppose that light consists of transverse electromagnetic disturbances.

On this assumption, the equations represent a beam of plane polarized light. They show that, in such a beam, there are at the same time electric and magnetic vibrations, both transverse, and at right angles to each other. It must be added that the electromagnetic field is the seat of two kinds of energy distinguished by the names of electric and magnetic energy, and that, according to a beautiful theorem due to J. H. Poynting, 4 the energy may be conceived to flow in a direction perpendicular both to the electric and to the magnetic force.

The amounts per unit of volume of the electric and the magnetic energy are given by the expressions,

2 (+ Ey Dy + Ez Dz),

a (HxBx +Hy B y+HzBz) = whose mean values for a full period are equal in every beam of light. The index of refraction of a body is given by result that has been verified by Ludwig Boltzmann's measurements' of the dielectric constants of gases. Thus Maxwell's theory can assign the true cause of the different optical properties of various transparent bodies.

It also leads to the reflection formulae, provided the electric vibrations of polarized light be supposed to be perpendicular to the plane of polarization, which implies that the magnetic vibrations are parallel to that plane. Following the same assumption Maxwell deduced the laws of double refraction, which he ascribes to the unequality of E l, 21 His results agree with those of Fresnel and the theory has been confirmed by Boltzmann,' who measured the three coefficients in the case of crystallized sulphur, and compared them with the principal indices of refraction. Subsequently the problem of crystalline reflection has been completely solved and it has been shown that, in a crystal, Poynting's flow of energy has the direction of the rays as determined by Huygens's construction.

Two further verifications must here be mentioned. In the first place, though we shall speak almost exclusively of the propagation of light in transparent dielectrics, a few words may be said about the optical properties of conductors. The simplest assumption concerning the electric current C in a metallic body is expressed by the equation C = QE, where a is the coefficient of conductivity.

Combining this with his other formulae. Maxwell found that there must be an absorption of light, a result that can be readily understood since the motion of electricity in a conductor gives rise to a development of heat. But, though Maxwell accounted in this way for the fundamental fact that metals are opaque bodies, there remained a wide divergence between the values of the coefficient of absorption as directly measured and as calculated from the electrical conductivity; but in 1903 it was shown by E. Hagen and H. Rubens that the agreement is very satisfactory in the case of the extreme infra-red rays.

In the second place, the electromagnetic theory requires that a surface struck by a beam of light shall experience a certain pressure. If the beam falls normally on a plane disk, the pressure is normal too; its total amount is given by c-1, if i 1, i2 and i 3 are the quantities of energy that are carried forward per unit of time by the incident, the reflected, and the transmitted light.

This result has been quantitatively verified by E. F. Nicholls and G. F. Hull.' Maxwell's predictions have been splendidly confirmed by the experiments of Heinrich Hertz' and others on electromagnetic waves; by diminishing the length of these to the utmost, some physicists have been able to reproduce with them all phenomena of reflection, refraction (single and double), interference, and polarization.

The wave-lengths observed in the aether now has Let the indices *p* and n relate to the two principal cases in which the incident (and, consequently, the reflected) light is polarized in the plane of incidence, or normally to it, and let positive directions *h* and *h* be chosen for the disturbance in the incident and for that in the reflected beam, in such a manner that, by a common rotation, h and the incident ray prolonged may be made to coincide with h and the reflected ray.

Mechanical Models of the Electromagnetic Medium

From the results already enumerated, a clear idea can be formed of the difficulties which were encountered in the older form of the wave-theory. Whereas, in Maxwell's theory, longitudinal vibrations are excluded *ab initio* by the solenoidal distribution of the electric current, the elastic-solid theory had to take them into account, unless, as was often done, one made them disappear by supposing them to have a very great velocity of propagation, so that the aether was considered to be practically incompressible. Even on this assumption, however, much in Fresnel's theory remained questionable. Thus George Green,' who was the first to apply the theory of elasticity in an unobjectionable manner, arrived on Fresnel's assumption at a formula for the reflection coefficient *A n* sensibly.

In the theory of double refraction the difficulties are no less serious. As a general rule there are in an anisotropic elastic solid three possible directions of vibration, at right angles to each other, for a given direction of the waves, but none of these lies in the wave-front. In order to make two of them do so and to find Fresnel's form for the wave-surface, new hypotheses are required. On Fresnel's assumption it is even necessary, as was observed by Green, to suppose that in the absence of all vibrations there is already a certain state of pressure in the medium.

If we adhere to Fresnel's assumption, it is indeed scarcely possible to construct an elastic model of the electromagnetic medium. It may be done, however, if the velocities of the particles in the model arc taken to represent the magnetic force H, which, of course, implies that the vibrations of the particles are parallel to the plane of polarization, and that the magnetic energy is represented by the kinetic energy in the model.

Considering further that, in the case of two bodies connected with each other, there is continuity of H in the electromagnetic system, and continuity of the velocity of the particles in the model, it becomes clear that the representation of H by that velocity must be on the same scale in all substances, so that, if E, *n*, *I* are the displacements of a particle and *g* a universal constant, we may write Hx =gol, Hy-got ' Hz=gat.

By this the magnetic energy per unit of volume becomes zg^2 *(OE)* 2 *(mi)* 2 and since this must be the kinetic energy of the elastic medium, the density of the latter must be taken equal to g^2, so that it must be the same in all substances.

Theories of MacCullagh

A theory of light in which the elastic aether has a uniform density, and in which the vibrations are supposed to be parallel to the plane of polarization, was developed by Franz Ernst Neumann, 2 who gave the first deduction of the formulas for crystalline reflection. Like Fresnel, he was, however, obliged to introduce some illegitimate assumptions and simplifications.

Here again Green indicated a more rigorous treatment. James MacCullagh 3 avoided this complication by simply assumi -ng an expression of the form for the potential energy.

He thus established a theory that is perfectly consistent in itself, and may be said to have foreshadowed the electromagnetic theory as regards the form of the equations for transparent bodies.

Lord Kelvin afterwards interpreted MacCullagh's assumption by supposing the only action which is called forth by a displacement to consist in certain couples acting on the elements of volume and proportional to the components If their rotation from the natural position.

He also showed 4 that this "rotational elasticity" can be produced by certain hidden rotations going on in the medium. We cannot dwell here upon other models that have been proposed, and most of which are of rather limited applicability. A mechanism of a more general kind ought, of course, to be adapted to what is known of the molecular constitution of bodies, and to the highly probable assumption of the perfect permeability for the aether of all ponderable matter, an assumption by which it has been possible to escape from one of the objections raised by Newton.

The possibility of a truly satisfactory model certainly cannot be denied. But it would, in all probability, be extremely complicated. For this reason many physicists rest content, as regards the free aether, with some such general form of the electromagnetic theory as has been sketched.

Optical Properties of Ponderable Bodies

Theory of Electrons

If we want to form an: Adequate representation of optical phenomena in ponderable bodies, the conceptions of the molecular and atomistic theories naturally suggest themselves. Already, in the elastic theory, it had been imagined that certain material particles are set vibrating by incident waves of light.

These particles had been supposed to be acted on by an elastic force by which they are drawn back towards their positions of equilibrium, so that they can perform free vibrations of their own, and by a resistance that can be represented by terms proportional to the velocity in the equations of motion, and may be physically understood if the vibrations are supposed to be converted in one way or another into a disorderly heat-motion.

In this way it had been found possible to explain the phenomena of dispersion and (selective) absorption, and the connexion between them (anomalous dispersion). These ideas have, been also embodied into the electromagnetic theory.

In its more recent development the extremely small, electrically charged particles, to which the name of " electrons "has been given, and which are supposed to exist in the interior of all bodies, are considered as forming the connecting links between aether and matter, and as determining by their arrangement and their motion all optical phenomena that are not confined to the free aether. It has thus become clear why the relations that had been established between optical and electrical properties have been found to hold only in some simple cases.

However, the general boundary conditions seem to require no alteration. For this reason it has been possible, for example, to establish a satisfactory theory of metallic reflection, though the propagation of light in the interior of a metal is only imperfectly understood.

One of the fundamental propositions of the theory of electrons is that an electron becomes a centre of radiation whenever its velocity changes either in direction or in magnitude.

Thus the production of Röntgen rays, regarded as consisting of very short and irregular electromagnetic impulses, is traced to the impacts of the electrons of the cathode-rays against the anticathode, and the lines of an emission spectrum indicate the existence in the radiating body of as many kinds of regular vibrations, the knowledge of which is the ultimate object of our investigations about the structure of the spectra.

The shifting of the lines caused, according to Doppler's law, by a motion of the source of light, may easily be accounted for, as only general principles are involved in the explanation. To a certain extent we can also elucidate the changes in the emission that are observed when the radiating source is exposed to external magnetic foooptics.

Various Kinds of Light-motion

If the disturbance is represented by P z = o, P = a cos *(nt-kx+f)*, *'P Z = a'cos (nt-+ f)*, so that the end of the vector P describes an ellipse in a plane perpendicular to the direction of propagation, the light is said to be elliptically, or in special cases circularly, polarized. Light of this kind can be dissolved in many different ways into plane polarized components.

There are cases in which plane waves must be elliptically or circularly polarized in order to show the simple propagation of phase that is expressed by formulae. Instances of this kind occur in bodies having the property of rotating the plane of polarization, either on account of their constitution, or under the influence of a magnetic field. For a given direction of the wave-front there are in general two kinds of elliptic vibrations, each having a definite form, orientation, and direction of motion, and a determinate velocity of propagation. All that has been said about Huygens's construction applies to these cases.

In a perfect spectroscope a sharp line would only be observed if an endless regular succession of simple harmonic vibrations were admitted into the instrument. In any other case the light will occupy a certain extent in the spectrum, and in order to determine its distribution we have to decompose into simple harmonic functions of the time the components of the disturbance, at a point of the slit for instance. This may be done by means of Fourier's theorem.

An extreme case is that of the unpolarized light emitted by incandescent solid bodies, consisting of disturbances whose variations are highly irregular, and giving a continuous spectrum. But even with what is commonly called homogeneous light, no perfectly sharp line will be seen. There is no source of light in which the vibrations of the particles remain for ever undisturbed, and a particle will never emit an endless succession of uninterrupted vibrations, but at best a series of vibrations whose form, phase and intensity are changed at irregular intervals. The result must be a broadening of the spectral line.

In cases of this kind one must distinguish between the velocity of propagation of the phase of regular vibrations and the velocity with which the said changes travel onward.

In a train of plane waves of definite frequency the disturbance is represented by means of goniometric functions of the time and the coordinates. Since the fundamental equations are linear, there are also solutions in which one or more of the coordinates occur in an exponential function. These solutions are of interest because the motions corresponding to them are widely different from those of which we have thus far spoken.

If, for example, the formulae contain the factor e-*zcos (*nt-sy +1) with the positive constant r, the disturbance is no longer periodic with respect to x, but steadily diminishes as x increases. A state of things of this kind, in which the vibrations rapidly die away as we leave the surface, exists in the air adjacent to the face of a glass prism by which a beam of light is totally reflected. It furnishes us an explanation of Newton's experiment mentioned.

Velocity of Light

The fact that light is propagated with a definite speed was first brought out by Ole Roemer at Paris, in 1676, through observations of the eclipses of Jupiter's satellites, made in different relative positions of the Earth and Jupiter in their respective orbits. It is possible in this way to determine the time required for light to pass across the orbit of the earth.

The dimensions of this orbit, or the distance of the sun, being taken as known, the actual speed of light could be computed. Since this computation requires a knowledge of the sun's distance, which has not yet been acquired with certainty, the actual speed is now determined by experiments made on the earth's surface.

Were it possible by any system of signals to compare with absolute precision the times at two different stations, the speed could be determined by finding how long was required for light to pass from one station to another at the greatest visible distance. But this is impracticable, because no natural agent is under our control by which a signal could be communicated with a greater velocity than that of light. It is therefore necessary to reflect a ray back to the point of observation and to determine the time which the light requires to go and come. Two systems have been devised for this purpose. One is that of Fizeau, in which the vital appliance is a rapidly revolving toothed wheel; the other is that of Foucault, in which the corresponding appliance is a mirror revolving on an axis in, or parallel to, its own plane.

The principle underlying Fizeau's method is shown in the accompanying the course of a ray of light which, emanating from a luminous point L, strikes the plane surface of a plate of glass M at an angle of about 45° A fraction of the light is reflected from the two surfaces of the glass to a distant reflector R, the plane of which is at right angles to the course of the ray. The latter is thus reflected back on its own course and, passing through the glass M on its return, reaches a point E behind the glass. An observer with his eye at E iooking through the glass sees the return ray as a distant luminous point in the reflector R, after the Nam light has passed over the course in both directions trance from M nearly equal to its focal length. The function of this appliance is to render the diverging rays, shown by the dotted lines, nearly parallel, in order that more light may reach R and be thrown back again. But the principle may be conceived without respect to the telescope, all the rays being ignored except the central one, which passes over the course we have described.

The luminous point between two of the teeth at K. Now, conceive that the wheel is set in revolution. The ray is then interrupted as every tooth passes, so that what is sent out is a succession of flashes. Conceive that the speed of the mirror is such that while the flash is going to the distant mirror and returning again, each tooth of the wheel takes the place of an opening between the teeth.

Then each flash sent out will, on its return, be intercepted by the adjacent tooth, and will therefore become invisible. If the speed be now doubled, so that the teeth pass at intervals equal to the time required for the light to go and come, each flash sent through an opening will return through the adjacent opening, and will therefore be seen with full brightness.

If the speed be continuously increased the result will be successive disappearances and reappearances of the light, according as a tooth is or is not interposed when the ray reaches the apparatus on its return. The computation of the time of passage and return is then very simple. The speed of the wheel being known, the number of teeth passing in one second can be computed. The order of the disappearance, or the number of teeth which have passed while the light is going and coming, being also determined in each case, the interval of time is computed by a simple formula.

The most elaborate determination yet 'made by Fizeau's method was that of Cornu. The station of observation was at the Paris Observatory. The distant reflector, a telescope with a reflector at its focus, was at Montlhery, distant 22,910 metres from the toothed wheel. Of the wheels most used one had 150 teeth, and was 35 millimetres in diameter; the other had 200 teeth, with a diameter of 45 mm. The highest speed attained was about 900 revolutions per second. At this speed, 135,000 (or 180,000) teeth would pass per second, and about 20 (or 28) would pass while the light was going and coming. But the actual speed attained was generally less than this.

The definitive result derived by Cornu from the entire series of experiments was 300,400 kilometres per second. Further details of this work need not be set forth because the method is in several ways deficient in precision. The eclipses and subsequent reappearances of the light taking place gradually, it is impossible to fix with entire precision upon the moment of complete eclipse. The speed of the wheel is continually varying, and it is impossible to determine with precision what it was at the instant of an eclipse.

The defect would be lessened were the speed of the toothed wheel placed under control of the observer who, by action in one direction or the other, could continually check or accelerate it, so as to keep the return point of light at the required phase of brightness. If the phase of complete extinction is chosen for this purpose a definite result cannot be reached; but by choosing the moment when the light is of a certain definite brightness, before or after an eclipse, the observer will know at each instant whether the speed should be accelerated or retarded, and can act accordingly.

The nearly constant speed through as long a period as is deemed necessary would then be found by dividing the entire number of revolutions of the wheel by the time through which the light was kept constant. But even with these improvements, which were not actually tried by Cornu, the estimate of the brightness on which the whole result depends would necessarily be uncertain. The outcome is that, although Cornu's discussion of his experiments is a model in the care taken to determine so far as practicable every source of error, his definitive result is shown by other determinations to have been too great by about a t°u part of its whole amount. In Foucault's determination the measures were not made upon a luminous point, but upon a reticule, the image of which could not be seen unless the reflector was quite near the revolving mirror. Indeed the whole apparatus was contained in his laboratory. The effective distance was increased by using several reflectors; but the entire course of the ray measured only 20 metres.

The result reached by Foucault for the velocity of light was 298,000 kilometres per second. The first marked advance on Foucault's determination was made by Albert A. Michelson, then a young officer on duty at the U.S. Naval Academy, Annapolis. The improvement *Michelson* consisted in using the image of a slit through which the rays of the sun passed after reflection from a heliostat. In this way it was found possible to see the image of the slit reflected from the distant mirror when the latter was nearly 600 metres from the station of observation.

It will be seen that the revolving mirror is here interposed between the lens and its focus. It was driven by an air turbine, the blast of which was under the control of the observer, so that it could be kept at any required speed. The speed was determined by the vibrations of two tuning forks. One of these was an electric fork, making about 120 vibrations per second, with which the mirror was kept in unison by a system of rays reflected from it and the fork. The speed of this fork was determined by comparison with a freely vibrating fork from time to time. The speed of the revolving mirror was generally about 275 turns per second, and the deflection of the image of the slit about 112.5 mm.

In order that the disturbances of the return image due to the passage of the ray through more than 7 km. of air might be reduced to a minimum, an ordinary telescope of the " broken back "form was used to send the ray to the revolving mirror. The speed of the mirror was, as in Michelson's experiments, completely under control of the observer, so that by drawing one or the other of two cords held in the hand the return image could be kept in any required position.

In making each measure the receiving telescope hereafter described was placed in a fixed position and during the "run" the image was kept as nearly as practicable upon a vertical thread passing through its focus. A “run“ generally lasted about two minutes, during which time the mirror commonly made between 25,000 and 30,000 revolutions. The speed per second was found by dividing the entire number of revolutions by the number of seconds in the “run”

The extreme deviations between the times of transmission of the light, as derived from any two runs, never approached to the thousandth part of its entire amount. The average deviation from the mean was indeed less than Pa 0 part of the whole. To avoid the injurious effect of the directly reflected flash, as well as to render unnecessary a comparison between the directions of the outgoing and the return ray. a second telescope, turning horizontally on an axis coincident with that of the revolving mirror, was used to receive the return ray after reflection.

This required the use of an elongated mirror of which the upper half of the surface reflected the outgoing ray, and the lower other half received and reflected the ray on its return. On this system it was not necessary to incline the mirror in order to avoid the direct reflection of the return ray.

The greatest advantage of this system was that the revolving mirror could be turned in either direction without break of continuity, so that the angular measures were made between the directions of the return ray after reflection when the mirror moved in opposite directions. In this way the speed of the mirror was as good as doubled, and the possible constant errors inherent in the reference to a fixed direction for the sending telescope were eliminated. The revolving mirror was a rectangular prism M of steel, 3 in. high and 12 in. on a side in cross section, which was driven A by a blast of air acting on two fanwheels, one at the top, the other at the bottom of the mirror. NPO is the object-end of the fixed sending telescope the rays passing through it’ being reflected to the mirror by a prism P.

The receiving telescope ABO is straight, and has its objective under O. It was attached to a frame which could turn around the same axis as the mirror. The angle through which it moved was measured by a divided arc immediately below its eye-piece. The position AB is that for receiving the ray during a rotation of the mirror in the anti-clockwise direction; the position A’B’ that for a clockwise rotation.

In these measures the observing station was at Fort Myer, on a hill above the west bank of the Potomac river. The distant reflector was first placed in the grounds of the Naval Observatory, at a distance of 2551 metres. But the definitive measures were made with the reflector at the base of the Washington monument, 3721 metres distant. The revolving mirror was of nickel-plated steel, polished on all four vertical sides.

Thus four reflections of the ray were received during each turn of the mirror, which would be coincident were the form of the mirror invariable. During the preliminary series of measures it was found that two images of the return ray were sometimes formed, which would result in two different conclusions as to the velocity of light, according as one or the other was observed. The only explanation of this defect which presented itself was a tortional vibration of the revolving mirror, coinciding in period with that of revolution, but it was first thought that the effect was only occasional.

In the summer of 1881 the distant reflector was removed from the Observatory to the Monument station. Six measures made in August and September showed a systematic deviation of +67 km. per second from the result of the Observatory series. This difference led to measures for eliminating the defect from which it was supposed to arise. The pivots of the mirror were reground, and a change made in the arrangement, which would permit of the effect of the vibration being determined and eliminated. This consisted in making the relative position of the sending and receiving telescopes interchangeable. In this way, if the measured deflection was too great in one position of the telescopes, it would be too small by an equal amount in the reverse position.

As a matter of fact, when the definitive measures were made, it was found that with the improved pivots the mean result was the same in the two positions. But the new result differed systematically from both the former ones. Thirteen measures were made from the Monument in the summer of 1882, the results of which will first be stated in the form of the time required by the ray to go and come.

So far as could be determined from the discordance of the separate measures, the mean error of Newcomb's result would be less than Da km. But making allowance for the various sources of systematic error the actual probable error was estimated at =30 km. It seems remarkable that since these determinations were made, a period during which great improvements have become possible in every part of the apparatus, no complete redetermination of this fundamental physical constant has been carried out. The experimental measures thus far cited have been primarily those of the velocity of light in air, the reduction to a vacuum being derived from theory alone.

The fundamental constant at the basis of the whole theory is the speed of light in a vacuum, such as the celestial spaces. The question of the relation between the velocity in vacuo, and in a transparent medium of any sort, belongs to the domain of physical optics.

Referring to the preceding section for the principles at play we shall in the present part of the article confine ourselves to the experimental results. With the theory of the effect of a transparent medium is associated that of the possible differences in the speed of light of different colours. The question whether the speed of light in vacuo varies with its wave-length seems to be settled with entire certainty by observations of variable stars. These are situated at different distances, some being so far that light must be several centuries in reaching us from them.

Were there any difference in the speed of light of various colours it would be shown by a change in the colour of the star as its light waxed and waned. The light of greatest speed preceding that of lesser speed would, when emanated during the rising phase, impress its own colour on that which it overtook.

The slower light would predominate during the falling phase. If there were a difference of Io minutes in the time at which light from the two ends of the visible spectrum arrived, it would be shown by this test. As not the slightest effect of the kind has ever been seen, it seems certain that the difference, if any, cannot imate to T. 0 0.0 0 0 part rt of the entire speed.

The approx case is different when light passes through a refracting medium. It is a theoretical result of the undulatory theory of light that its velocity in such a medium is inversely proportional to the refractive index of the medium. This being different for different colours, we must expects a corresponding difference in the velocity. Foucault and Michelson have tested these results of the undulatory theory by comparing the time required for a ray of light to pass through a tube filled with a refracting medium, and through air. Foucault thus found, in a general way, that there actually was a retardation; but his observations took account only of the mean retardation of light of all the wavelengths, which he found to correspond with the undulatory theory. Michelson went further by determining the retardation of light of various wave-lengths in carbon bisulphide.

The comparison of red and blue light was made differentially. The colours selected were of wave-length about offoered and 0.49 for blue. Putting Vr and Vb for the speeds of red and blue light respectively in bisulphide of carbon, the mean result compares with theory as follows: Observed value of the ratio V r, Theoretical value (Verdet). 1.025 This agreement may be regarded as perfect. It shows that the divergence of the speed of yellow light in the medium from theory, as found above, holds through the entirespectrum. The excess of the retardation above that resulting from theory is probably due to a difference between " wave-speed" and "group-speed" pointed out by Rayleigh. But when a flash of light like that measured passes through a refracting medium, the front waves of the flash are continually dying away, and the place of each is taken by the wave following. A familiar case of this sort is seen when a stone is thrown into a pond. The front waves die out one at a time, to be followed by others, each of which goes further than its predecessor, while new waves are formed in the rear.

Larger waves in the middle, moves as a whole more slowly than do the individual waves. When the speed of light is measured the result is not the wave-speed as above defined, but something less, because the result depends on the time of the group passing through the medium. This lower speed is called the groupvelocity of light. In a vacuum there is no dying out of the waves, so that the group-speed and the wave-speed are identical.

Chapter 2

Variational Methods

Properties of Solids

Solid and Liquid Phases

The vast majority of known substances are solids under ordinary conditions of pressure and temperature. An understanding of the structures of crystalline solids at the atomic/molecular level. Be forewarned that a study of such structures requires a sound basis in geometry and an ability to visualize in 3-dimensions.

Further, it is impossible to obtain an adequate understanding of the important structural ideas from text and 2-dimensional figures. Work with 3-dimensional models is essential. Understand it as best you can, then refine your understanding with models in the laboratory. A solid base for the understanding of two major classes of materials: metals, in which all atoms in the structure are the same; and ionic compounds, composed of cations of one substance and anions of another.

Macroscopic Properties of Solids

Solids have a number of observable properties in common:

- They are rigid, maintaining fixed shape and volume independent of container.
- The molar volumes of solids are small.
- Solids are incompressible.
- Solids do not flow, and diffusion in solids is extremely slow.
- Many solids occur as crystals — well-defined shapes with structures suggesting an orderly internal arrangement of atoms, molecules, or ions.

To rationalize these properties we need a simple model of the solid phase.

In designing such a model, we make use of two facts:

1. Since the molar volume of a solid is much less than the molar volume of the gas phase of a substance, the molecules must be crowded together in the solid;
2. The orderly shapes of crystals suggests an orderly arrangement of particles. A simple model, then, is that solids consist of tightly-packed orderly arrangements of particles. Figure shows a 2-dimensional representation of a solid.

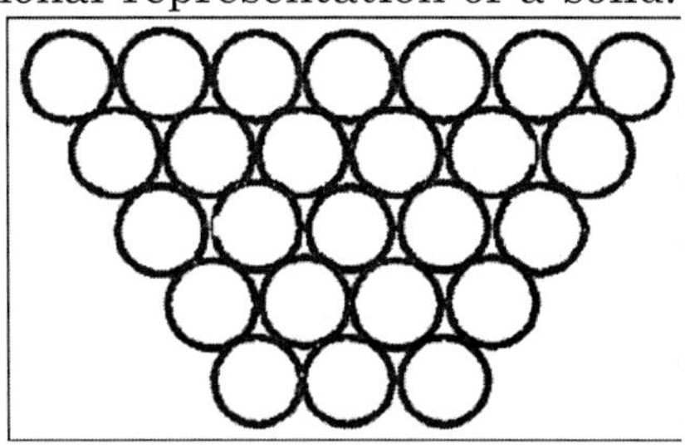

Figure: 2-dimensional representation of a solid

Note first how close together the particles are compared with the gas phase. This crowding allows very little free volume to the molecules, which strongly resist further compression. The solid is incompressible. It is also clear why diffusion is slow in solids, because each particle is locked in place by the surrounding ones.

Only at the surface of the solid, where a particle has neighbours on only 3 sides, can diffusion occur. The particles are in constant motion, just as in the gas phase. A molecule is constantly jiggling in place, bumping into neighbours that hold it in a cage. Just as for gases, the kinetic energies of the particles (atoms, molecules, or ions) of the solid are distributed according to the Maxwell-Boltzmann distribution, with the kinetic energy of an average particle having the value 3kT/2. The major difference between the motion in gas and solid is that in the solid, a particle is hemmed in and undergoes many more collisions per unit time than in the gas. The movement of a particle in the solid is much like a vibration, because the molecule stays in the same location. If the temperature of the solid is increased, the particles vibrate faster, but are still not free to move about because intermolecular forces from the neighbours hold them in place. Finally, the arrangement of articles is very orderly, not at all chaotic like the gas.

Each particle is surrounded by a perfect hexagon of other particles; and particles are arranged in rows that remain straight over long distances. It is easy to imagine that this orderly arrangement at the molecular level is manifested in beautiful crystalline shapes at the macroscopic level.

The Lattice Concept

We will describe structures of crystalline solids in terms of the lattice. A lattice is a regular, repeating pattern of points in space. Wallpaper patterns are examples of 2-dimensional lattices. The lattice is more fundamental than the particular wallpaper pattern shown, because wallpaper patterns using many different design units could be generated from the same lattice. Figure shows a second small collection of points that might be considered as a unit cell. Movement ofthis collection in directions parallel to its sides does indeed reproduce the entire lattice.

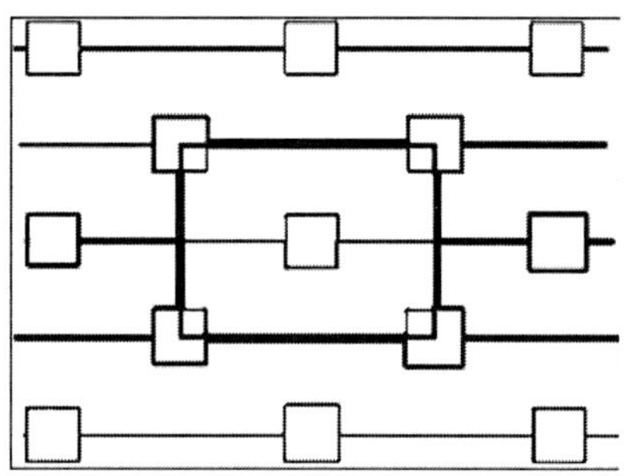

The key feature of the lattice is the arrangement of points. A small collection of lattice points that contains all of the spatial information in the lattice pattern is called a unit cell. The unit cell represents the lattice, because the whole lattice can be reproduced from the unit cellby moving (translating) the unit cell in directions parallel to its sides by distances equal to the lengths of its sides. However, the unit cell in Figure is simpler than that in Figure because it involves fewer points — one point at each of its 4 corners. A unit cell having lattice points at its corners only is called a primitive unit cell. The unit cell in Figure is a non-primitive unit cell because it has an additional point at its centre. If it is possible to represent a lattice with a primitive unit cell, we will do so.

Crystalline solids are represented by 3-dimensional lattices, one example of which is the simple cubic lattice. This is a primitive unit cell because it has points only at the corners. Movement of the unit cell by the distance, d, in directions parallel to its 3 dimensions reproduces the entire simple cubic lattice. The simple cubic lattice is analogous to an orderly arrangement of building blocks stacked face to face in 3-dimensions. It was shown in the 1800s by Bravais that there are only 7 shapes (called crystal types) that a unit cell can have and only 14 possible distributions of points (lattice types) within these 7 shapes. The 7 shapes, characterized by the relative magnitudes of the repeat distances in three directions and the angles between these directions, are summarized in Table.

Table: The Bravais Crystal Types

System	Edge Lengths	Angles
Cubic	a = b = c	all 90°
Tetragonal	a = b not equal to c	all 90°
Orthorhombic	a not equal to b not equal to c	all 90°
Monoclinic (once-angled)	a not equal to b not equal to c	a = b = 90°, g not equal to 90°
triclinic (thrice-angled)	a not equal to b not equal to c	a not equal to b not equal to g not equal to 90°
Rhombohedral	a = b = c	= 90°
Hexagonal	a = b not equal to c	a = b = 90°; g = 120°

We will discuss cubic and hexagonal lattices in the context of a discussion of close-packing ideas.

Close Packing

What is the most efficient way of arranging spheres in 3-dimensional space? We take "efficient" to mean economical in terms of space. We begin by building one layer of spheres that are arranged as closely to one another as possible. Note that this arrangement is identical to that pictured earlier.

Spheres are lined up in straight rows, with the spheres in one row fitting snugly into the indentations in adjacent rows. Each sphere is surrounded by a perfect hexagon of spheres, all touching it. There are numerous gaps, or holes, in the layer, of triangular shape. For descriptive purposes here, we label these holes "u" if the triangle has a single vertex pointing up, or "d" if the triangle has vertex down.

We now add a single sphere ofa second layer on to of the first. The added sphere rests in a low point of the first layer, corresponding with a triangular hole, in contact with 3 spheres of the first layer.

The 4 touching spheres lie at the vertices of a regular tetrahedron. There remains some unfilled space at the centre of this 4-sphere group that is called a tetrahedral hole. Each time we add a sphere to the second layer, we create a tetrahedral hole between the layers. When additional spheres are added to the second layer, it turns out that they must all go in either “u” holes or “d” holes, but not both.

Thus only half of the triangular depressions in the first layer can be used for placement of 2nd-layer spheres. Triangular holes in the 2nd layer lie directly over the unused holes of the first layer, so that channels pass right through both layers.

The point at the centre of such a channel is surrounded by 6 spheres, arranged as 2 equilateral triangles, one triangle in the 1st and one in the 2nd layer. These 6 spheres form a regular octahedron, a geometrical solid with 6 vertices and 8 equilateral triangular faces.

The space surrounded by the 6 spheres is called an octahedral hole. An octahedral hole is larger than a tetrahedral hole because 6 spheres cannot be crowded as closely together as 4 spheres.There are 2 ways in which a third layer of spheres can be added to the existing 2 layers in close-packed fashion. In the first way, spheres are place directly over spheres of the first layer. The third and first layers then have spheres in the same locations.

The 4th layer then goes over the second, the 5th over the 3rd, and so on, so a pattern that repeats every other layer is obtained. This is characterized as a 1212... pattern. The unit cell is hexagonal in shape; the 1212... close-paking arrangement is therefore called hexagonal close packing (hcp). Note that the coodination number (the number of nearest neighbours) of any sphere in hcp is 12.

The second way of adding a third layer is to place spheres directly over the octahedral holesbetween the first two layers. In this case, the third-layer spheres do not lie over spheres in layer 1. This is still a close-packing arrangement because the coordination number of any sphere is 12, just as in hcp. The 4th layer of spheres is then placed directly over the spheres in the first layer.

This type of close packing is designated 123123... to indicate that the pattern repeats after 3 layers. The unit cell for 123... close packing. Although it may not be obvious from the figure, this unit cell is a cube (this is a situation in which a 2-dimensional drawing does not adequately convey what is true). This type of close packing is therefore called cubic close packing (ccp). Hcp and ccp are the two ways of close packing spheres in 3 dimensins.

They have several features in common:

- Spheres occupy 74 per cent of the available space;
- Each sphere has 12 nearest neighbours that touch it (coordination number);
- The number of octahedral holes in the lattice is exactly the same as the number of spheres in the lattice, and there are twice as many tetrahedral holes as spheres.

There are a numbr of substances that crystallize in one or the other of these two packing arrangements. Beryllium, magnesium, and zinc metal all form hexagonal close packed lattices. Nickel, copper, gold, and aluminum form ccp lattices. Cubic close packing is particularly common. We will take a closer look at cubic lattices now. The three Bravais cubic lattices. The simple cubic (sc) lattice is primitive, with points (spheres or atoms) only at the vertices of the cube. Metallic polonium and non-metallic dioxygen (O_2) crystallize with the simple cubic lattice. The coordination number of an atom in the simple cubic lattice is 6. To determine this, pick one atom in the unit cell; find an atom that appears to be closest to the atom you picked; then count the number of other atoms that are at this same distance from the atom you picked.

In counting coordination numbers, you may have to go outside of the single unit cell. In the simple cubic lattice, identical unit cells surround the pictured one on all sides. Some of the nearest neighbours of your picked atom lie in these neighbouring unit cells. Finally, the simple cubic unit cell contains only 1 atom, despite the fact that you see 8 atoms drawn in the figure. We say that the simple cube has 1 atom per unit cell (1 atom/uc). The procedure for counting atoms in unit cells is described in the paragraph below. The body-centered cubic (bcc) lattice is non-primitive, with points at the corners and the centre of the unit cell (the centre of a cube i called the cube body centre). Metallic iron and barim crystallize in the bcc lattice.

The coordination number is 8 for this lattice (count the nearest neighbours of the body-centre atom), and the body-centered cube has 2 atoms per unit cell. Finally, the face-centered cubic (fcc) lattice is also non-primitive with points at each corner and in the centre of each of the 6 faces of the cube. Metallic nickel and copper adopt this structure. The coordination number is 12 for this lattice, which means that it must be close packed! The face-centered cubic lattice and the cubic close packed lattice are identical. It is very difficult to see this in two-dimensional drawings. To see it, you must imagine looking along a body diagonal of the face-centered cube. A body diagonal runs on a straight line from one corner, through the body centre of the cube, to the diagonally opposite corner. As you look along the body diagonal, you will see the pattern of spheres: a single sphere; a triangle consisting of 6 spheres; a second triangle of 6 spheres, but rotated 60° with respect to the first one; and finally another single sphere immediately below the first one. The cube is admittedly difficult to visualize from the drawings. The face-centered cubic lattice has 4 atoms per unit cell.

We now address the problem of determining the number of atoms per unit cell. This is quite simple once it is realised that atoms are generally shared among several unit cells and therefore must be partitioned accordingly. An atom at a corner (vertex) of a cube is shared among 8 unit cells. To see this, imagine the simple cube in Figure to be a building block, and focus on the upper right front corner of the building block. Imagine that this corner is labelled "a".

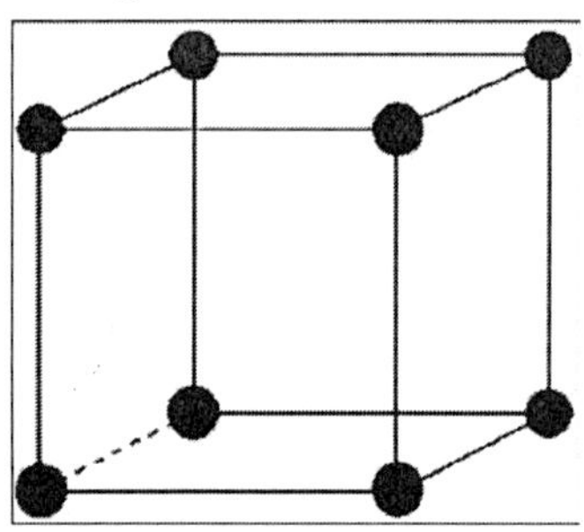

In your mind, place identical blocks on top, to the right, and in front of the block in the figure. Then place additional blocks in the remaining places adjoining these blocks to generate the 8-block.

Point "a" is now at the centre of this collection, and is simultaneously a corner for each of the 8 blocks; 8 blocks share a vertex. This point cannot be said to belong entirely to any particular block. Instead, each block gets 1/8 of it. The counting rule for a corner atom, then, is that only 1/8 of it is assigned to a single unit cell.

Similar thinking leads to the conclusion that a face-centre atom is shared between 2 unit cells, so 1/2 of it is assigned to a particular unit cell; and a body-centre atom is not shared at all, so it belongs entirely to the unit cell that it is found in. The atom count for a body-centered unit cell then goes like this:

atoms/uc = 8 corner atoms x 1/8 + 1 body-centre atom x 1

We will have occasion also to partition points located along edges of the cube. Edges are the lines connecting adjacent vertices. Since an edge is common to 4 cubes arranged in a square, 1/4 of any point along an edge is assigned to a particular cube.

Table: Atom Counting Rules for Cubic Unit Cells

Atom location	*Fraction Assigned to uc*
Corner	1/8
Edge	1/4
Face	1/2
Body-center	1

If, as we have said, the face-centered cubic lattice is in fact cubic close packed, then there should be tetrahedral and octahedral holes in the fcc lattice. Further, we should find 4 octahedral and 8 tetrahedral holes in a fcc unit cell based on the the last-listed common feature of close-packed lattices above. To locate these holes it is convenient to place our cube in a coordinate system.

The origin of coordinates is located at the lower left front corner of the cube, with the x-axis along the bottom front edge, the y-axis along the left vertical edge, and the z axis along the left bottom edge.

We specify distances along the 3 axes in terms of fractions of the cube edge length. The set of coordinates (1,0,0) means to proceed 1 edge length along x. These are the coordinates of the lower right front vertex of the cube (labelled 1 in the figure). Similarly, (1/2,1/2,1/2) specifies the body centre of the cube, labelled 2 in the figure.

Now examine the face-centered cube. We find the most obvious octahedral hole at the body centre. This point is surrounded by 6 atoms all at the same distance from it. These are the atoms at the face centers of the cube. Similar octahedral holes are located at the cube edge centers. It is surrounded at equal distance by the atoms at (1/2,1/2,1), (1,0,1), (1,1,1), and (1,1/2,1/2). The remaining 2 atoms are not part of the unit cell shown, but can still be located with coordinate sets (1 1/2,1/2,1) and (1,1/2,-1 1/2). (Translating these coordinates is good practice in the use of the unit cell coordinate system.)

The cube has 12 edges, so there are 12 holes of this type. The counting rules tell us that only 1/4 of each of these edge-centered octahedral holes belongs to the unit cell of interest. The total number of octahedral holes in the unit cell is then 1(body centre) + 12(1/4) = 4. The number of octahedral holes is the same as the number of atoms in the unit cell! A tetrahedral hole in the fcc unit cell. It is defined by the four shaded atoms, and has coordinates (1/4,1/4,1/4). There are 7 other tetrahedral holes in the unit cell, all with coordinate combinations involving 1/4 and 3/4, and all defined by 1 corner and 3 face-centre atoms. Try to locate them in the figure, and write down their coordinates. Each tetrahedral hole lies entirely within the cube body, so none of them are shared with other unit cells. The number of tetrahedral holes per fcc unit cell is therefore 8, twice the number of atoms in the unit cell.

Octahedral and tetrahedral holes are important in the discussion of ionic compounds.

Ionic Compounds

The ideas discussed thus far apply to crystals in which the units — spheres or atoms — are all identical. Crystals of pure metals, with close-packed metal atoms, and even crystals of molecular solids (for example, methane, CH_4, in which there is a methane molecule at each lattice point) are examples of real situations in which these ideas apply. Ionic compounds are different because there are two distinctly different units — the cation and the anion — that must be accommodated in the lattice. The resulting lattices are often called compound lattices because they contain two different components. The cation and the anion differ in charge, of course, and in chemical identity (usually). They also differ in size. Generally speaking, cations are smaller than anions, for easily-understood reasons. Cations contain an excess of protons over electrons. The Coulomb pull of the excess positive charge causes the electron cloud to contract, so that it is smaller than in the corresponding atom (*e.g.*, Na^+ is smaller than Na).

Anions contain an excess of electrons over protons. The repulsion between the electrons exceeds that in the corresponding neutral atom, causing an expansion of the electron cloud (*e.g.*, Cl^- is larger than Cl). Except for a few rare cases, the cation in an ionic compound is smaller than the anion. In forming a lattice, the space requirements are more important for the larger anion, which tends to dictate which lattice is adopted.

The (smaller) cations then fit into the holes, or interstices, of the anion lattice. Many common ionic compounds have lattices that consist of cubic close packed anions, with cations in tetrahedral and/or octahedral holes. Stoichiometries of up to 3 cations per anion can be accommodated in this framework. We now discuss some common ionic lattices. The Sodium Chloride (Rock Salt) Lattice. It can be described as a cubic close packed (face-centered cubic) anion lattice, with Na^+ ions in the octahedral holes. The coordination number (number of nearest neighbours of opposite charge) of each Na^+ ion is 6, and of each Cl- ion is 6. The cation and anion must have equal coordination numbers because the stoichiometry of the compound is 1:1 (1 cation per anion).

The unit cell contains 4 formula units of NaCl (4 Cl- ions, which define the fcc unit cell; 4 Na^+ ions because there are 4 octahedral holes per unit cell). Many ionic compounds exhibit the NaCl lattice, including most of the group 1 halides, most of the group 2 oxides, and a number of d-metal oxides.

The Zinc Blende Lattice. Zinc blende is the common name for one crystal modification of the common mineral, ZnS. The other form is known as wurtzite. Wurtzite consists of an hcp array of S^{2-} ions with Zn^{2+} ions in one-half of the tetrahedral holes; it will not concern us further here, because we are restricting our attention to cubic lattices. Zinc blende consists of a ccp array of S^{2-}, with Zn^{2+} in half of the tetrahedral holes.

Since there are 8 tetrahedral holes in the sulfide ion lattice, but only 4 S^{2-} ions, only half of the holes are used to accommodate 4 Zn^{2+} ions. This is consistent with the 1:1 stoichiometry of the compound.

The zinc ions occupy tetrahedral holes so that they can be as far apart from one another as possible to minimize repulsions between positive ions. If the first Zn^{2+} ion occupies the hole at (1/4,3/4,1/4), the other three will be in holes at (3/4,1/4,1/4), (1/4,1/4,3/4), and (3/4,3/4,3/4). You should locate these holes and verify that the Zn^{2+} ions describe a tetrahedron. The cation and anion coordination numbers are both 4 in zincblende. As for NaCl, there are 4 formula units per unit cell. Other ionic compounds that adopt the zinc blende lattice include BeS, CdS, HgS, and AgI.

The Calcium Fluoride (fluorite) Lattice. This compound has formula CaF_2,and exhibits. The Ca^{2+} ion is virtually the same size as the F- ion, one of those rare situations referred to above, and forms a face-centered cubic lattice. The F- ions fill all of the tetrahedral holes in the cation lattice. Since there are 8 such holes per 4 Ca^{2+} ions, the stoichiometry is nicely accommodated.

The coordination umber of the Ca^{2+} ion is 8, and that of the - ion is 4. (Note that the number of cations per formula unit multiplied by the cation coordination number is equal to the product of the number of anions per formula unit and the anion coordination number.) There are again 4 formula units per unit cell. The fluorite structure is very common for ionic compounds of 1:2 (or 2:1) stoichiometry.

The Cryolite Lattice. Cryolite is the common name fr the mineral, a_3AlF_6. Though rare, this substance is of extreme commercial value, because it is used in the process of extraction of aluminum from bauxite ore. Cryolite consists of Na^+ cations and AlF_6^{3-} anions. It adopts a lattice in which the anions are cubic close packed (fcc) and the cations occupy all of the octahedal and tetrahedral holes. The Diamond Lattice. Diamond is on crystal modification of the element carbon. It is, of course, not ionic, nor ven a compound. It is dealt with here because it, too, is based on a fcc unit cell.

The diamond lattice is just like the zinc blend lattice, with carbon atoms in place of both Zn^{2+} and S^{2-} ions. We therefore describe it as a fcc array of carbon atoms with carbon atoms in one-half the tetrahedral holes. The coordination number of eac carbon atom in diamond is 4.

The Cesium Chloride Lattice. The cesium chloride lattice. It is a more open lattice than those considered thus far. This lattice consists of a simple cubic array of chloride ions, with Cs^+ ions in cubic holes. A cubic hole is the space at the body centre of a simple cubic unit cell. There is 1 formula unit of CsCl per unit cell, and the coordination number of both cation and anion is 8. Ionic compounds that adopt the CsCl lattice include CsBr CsI, the halides of thallium(I), and NH_4Cl.

Radius Ratio Rules. Compounds of 1:1 stoichiometry commony adopt one of three lattices discussed above: the zinc blende lattice, the rock salt lattice, or the cesium chloride lattice. Of interest now is whether or not we can predict, or at least understand, why a particular compound favours one lattice over the others. A number of factors are important in choice of lattice, but one of these dominates in importance: the relative sizes of cation and anion.

Because we view the cation as occupying holes in the anion lattice, we begin by calculating the sizes of the various hole types in relation to the size of the anion involved. The most straightforward calculation is for the octahedral hole. The figure shows one face of the fcc unit cell, with the anions just touching along a face diagonal of the cube. There is an octahedral hole at the centre of each of the 4 cube edges shown; we focus on the right edge.

The size of the octahedral hole will be taken as the distance from the edge centre to, say, the top right anion, along the cube edge. (The distances from the hole to the bottom right anion and to the face-centered anion are the same.) We can express the right edge length as a sum of anion and hole radi as in equation:$a = 2r^- + 2r^+$

Here r^- is the anion radius, r^+ the hole radius. (The + designation is used in anticipation of putting a cation in the hole.) Similarly, we express the length of the face diagonal (fd) in terms of anion radii. This is easy, since the anions touch along this line:

$$fd = 4r^-$$

Finally, we relate the fd and edge length va the Pythagoren Theorem:

$$(fd)^2 = 2a^2$$

Take the square root of both sides of 7-4-3:

$$fd = a(2)^{1/2}$$

Substitute 7-4-4 in 7-4-2 and rearrange for a:

$$a = 2(2)^{1/2}\ r^-$$

Substitute for a in 7-4-1 and solve for r^+:

$$r^+ = ((2)^{1/2}-1)\ r^- = 0.414\ r^-$$

The octahedral hole is somewhat less than half as large as the anions that create the hole. By similar geometric methods, equations for tetrahedral and cubic holes respectively, can be developed:

$$r^+ = 0.224\ r^-\ r^+ = 0.732\ r^-$$

You should try to obtain these expressions, making use of right triangles and Pythagorean relationships. (In addition to equation, which relates the face diagonal and edge length of a cube, equation relates the cube body diagonal (bd) to the edge length:

$$bd = (3)^{1/2}$$

A tetrahedral hole is about 1/2 the size of an octahedral hole and about 1/3 the size of a cubic hole. Only cations that are relatively small compared with the anion can be accommodated in tetrahedral holes. As cation size increases relative to anion size, the cation will prefer a larger hole, with a larger coordination number of anions.

Equations are the basis for the following radius-ratio rules:

- If the ratio of the cation and anion radii (r^+/r^-) lies in the range between 0.224 and 0.414, the cation prefers to occupy a tetrahedral hole.

- If r^+/r^- lies in the range between 0.414 and 0.732, the cation prefers an octahedral hole.
- If r^+/r^- exceeds 0.732, the cation prefers a cubic hole.

These rules make good sense. If $r^+/r^- < 0.224$, the cation is too small for the tetrahedral hole. The anions will be touching, but will not be touching cations. Since it is the cation-anion attractions that stabilize the lattice, whereas anion-anion or cation-cation repulsions tend to destabilize it, this is not a favourable situation. When $r^+/r^- = 0.224$, the cation just fits the tetrahedral hole. Although anions are still touching, they now touch cations as well. This is a stable arrangement. As r^+/r^- becomes > 0.224, the cation is too big for the tetrahedral hole and starts to push the anions apart.

The cation-anion attractions are maintained, because cations and anions still touch, but now anion-anion repulsions are decreased because the anions no longer touch. This is an even better arrangement than the perfect fit.

This situation continues to improve until r^+/r^- becomes equal to 0.414. At this value, the cation prefers to occupy an octahedral hole because it can increase its coordination number without increasing anion-anion repulsion to any significant extent. The number of cation-anion attractive interactions increases from 4 to 6, giving a more stable lattice. As r^+/r^- increases between 0.414 and 0.732, cations in octahedral holes maintain their contact with 6 anions while pushing the anions further and further apart, decreasing repulsions between them. Throughout this range of r^+/r^-, occupation of an octahedral hole becomes increasingly favourable until the radius ratio becomes 0.732. The cation can now increase its coordination number to 8 anions, again increasing the number of attractive cation-anion interactions. The radius ratio rules are summarized in Table. Radii for common cations and anions are given in Table.

Table: Radius Ratio Rules

Range of r^+/r^-	*Lattice Type*	*Example*
0.224-0.414	ZnS	ZnS
0.414-0.732	NaCl	KCl
> 0.732	CsCl	CsCl

Ion radii and the radius ratio rules can be used to predict the probable choice of lattice for an ionic compound.

Example. The radii of the Na^+ ion and the Br^- ion are 116 and 182 pm, respectively. Which of the three 1:1 lattices will be adopted by NaBr?

Solution. Calculate r^+/r^-:

$r^+/r^- = 116/182 = 0.64$

This falls between 0.414 and 0.732. The cation will prefer the octahedral hole. NaBr should therefore have the rock salt structure. It does.

The radius ratio rules provide useful guidelines to the lattice structures of ionic compounds. It is important to realise, however, that they are not hard and fast; the predictions made using the rules are often wrong. For example, in the zincblende form of ZnS, Zn^{2+} ions occupy tetrahedral holes in a fcc lattice of S^{2-} ions. In fact, this compound is the prototype for this type of structure.

However, the radius ratio rules predict that the Zn^{2+} ions should occupy octahedral holes! The radius ratio rules fail frequently because the relative size of the ions is not the only factor that is important in determining structure. Of particular importance is the tendency towards at least some covalent bonding between cation and anion. Covalency involves directional requirements, and may influence the type of hole preferred by the cation.

X-Ray Crystallography

In 1912, von Laue predicted that crystals could serve as diffraction gratings if the wavelength of radiation used were approximately the same as the distance between planes of atoms in the crystals. This spacing was known to be on the order of 1×10^{-10} m, corresponding to wavelengths in the x-ray region of the electromagnetic spectrum. Subsequently, WH Bragg discovered that when X-rays of appropriate wavelength are directed onto the smooth face of a crystal, the X-rays are diffracted and a pattern characteristic of the lattice structure of the crystal is obtained.

This observation is the foundation for the modern techniue of X-ray Crystallography, the most powerful method that we have for determining molecular structure. Like the types of spectroscopy, x-ray crystallography is based on the interaction of light with matter. Unlike spectroscopy, however, the light is not absorbed by the matter in this case. Instead, it is scattered from the atoms of the crystal in a very definite way. Analysis of the scattering enables the arrangements of atoms within the crystal to be deduced. The analysis of x-ray diffraction patterns is complex, even for materials of simple structure. However, thanks to the efforts of many crystallographers over many years, it is today done quite quickl and routinely.

Once a suitable crystal of a substanc is grown, the modern crystallographer is able to mount it in a fully automated x-ray diffractometer, the instrument that is usedfor the x-ray diffraction study. The diffractometer automatically obtains all of the necessary data for the dtermination of the lattice structure.

The data are then fed to one of a number of computer programmes that have been developed and refined for the analysis of such data, and the structure is "solved."

The output of the programme reveals not only the lattice structure, but also the locations of all of the atoms of a molecule of the substance. Structures of substances ranging from simple salts or molecular substances to huge complex proteins and enzymes can be determined using this approach.

The structures of two molecules, obtained from x-ray diffraction data. The arrangement of atoms in the relatively simple molecule, phthalocyanine.

Example. Aluminum crystallizes in a cubic lattice. X-ray diffraction reveals that the edge length of the unit cell is 405.0 pm. The density of Al is 2.7 g/cm^3. Which cubic lattice does Al adopt?

Solution. We adopt the following strategy:

unit cell mass $\rightarrow$ Number Al atoms per uc $\rightarrow$ lattice type

We can obtain the unit cell mass from the density by calculating unit cell volume from the cube edge length.

Volume of uc = $a^3 = (4.05\times10^{-8}\text{ cm})^3 = 6.643\times10^{-23}$ cm^3/uc

Mass of unit cell = r x Volume uc = 2.7 g/cm$^3\times6.643\times10^{-23}$ cm^3/ uc = 1.794×10^{-22} g

Moles Al per uc = mass/molar mass = 6.647×10^{-24} moles AlAtoms Al per uc = 6.647×10^{-24} moles$\times6.022\times10^{23}$ atoms/mol= 4.0

The unit cell is face-centered cubic. This example shows the connection between macroscopic and microscopic properties. Note that in a reversal of the above procedure, knowledge of the type of cubic lattice could be used to experimentally measure Avogadro's number, N_o.

Example. In early nuclear reactors, fuel rods were made of uranium metal, with a cubic close packed lattice. These suffered from the problm that the fuel rods frequently burst out of their protective sheaths, which could obviously cause major safety problems. Why did this occur, and what could be done to prevent it?

Solution. Nuclear fission (the spontaneous breakup of a radioactive nucleus) produces 2 nuclei for each one that decays. One possible decay route is shown below:

$$^{235}U \rightarrow {}^{90}Sr + {}^{143}Xe + 2\ {}^{1}n$$

In the ccp U lattice, the holes are too small to accommodate these extra nuclei. The resulting expansion of the lattice caused bursting of the protective sheaths. The solution of the problem was to make the fuel rods from UO_2 (yellow cake), which consists of ccp U atoms with O atoms in the tetrahedral sites. The larger octahedral holes in this lattice were able to accommodate the fission fragments without expansion.

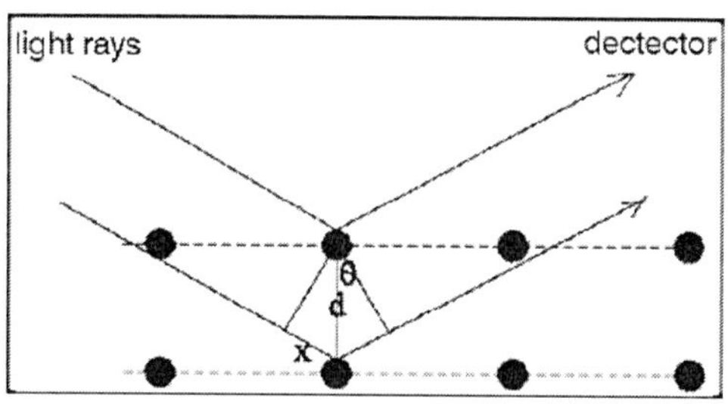

The Bragg Equation. Figure shows two planes of atoms in a crystal. A source of X-rays at the left emits radiation, two beams of which are shown impinging on the planes of atoms and scattering off to the right, where they are "seen" at a detector.

The angles made by the incident beam and the deflected beam with the planes of atoms are the same. A diffraction pattern is a series of alternating light and dark lines (sometimes spots). The white spots result from the arrival at the detector of reflected beams that are all in-phase. That is, their waves are matched up peak for peak and trough for trough. Waves that are in phase add together to give a wave with larger peaks and larger troughs — *i.e.*, more intensity. This is called constructive interference.

The white spots in the diffraction pattern correspond to deflection angles that give in-phase waves. Similarly, the dark areas correspond to angles that give out-of-phase waves. Here the peak of one wave arrives at the detector at the same time as a trough of a second wave arrives. Such waves cancel each other, and very little or no signal is seen at the detector. This is called destructive interference.

The criterion that must be met in order that two waves arrive at the detector in phase is that the difference in the distances travelled by them must correspond to an integer number of wavelengths. This will give the maximum intensity at the detector. If the difference in distance travelled corresponds to a half-integral number of wavelengths, cancellation will occur, and there will be no intensity registered at the detector. The difference in the distances travelled by the two deflected beams in Figure above is 2x. Since x is related to the angle of incidence (deflection) and to the separation of the planes of scattering atoms as x = dsinq, and since 2x must be an integer number, n, of wavelengths in order to maximize intensity, equation results.

$$n = 2d\sin q$$

This is known as the Bragg equation. It is the basis of modern x-ray crystallography. It shows that if we measure the angles at which constructive interference occurs, we can calculate d, the spacing between atom planes in the crystal.

Example. A set of planes in a crystal is separated by 250 pm (2.50×10^{-10} m). What is the maximum x-ray wavelength that could be used to measure this distance?

Solution. The largest scattering angle possible is 90°. Since sin $90^o = 1$, n = 2d = 5.00×10^{-10} m. q is largest for n = 1, giving = 5.00×10^{-10} m. A wavelength longer than this cannot fit even once into the distance, 2d, so cannot be used to measure the distance. X-rays with less than 5.00×10^{-10} m must be used.

Lattice Imperfections

Perfect crystals, in which every atom or molecule is in place, are an idealization, like the ideal gas. Real crystals contain fairly large numbers of imperfections, called defects. Three major types of defect are called point defects, line defects, and plane defects. Common point defects in simple lattices (in which all lattice points are identical) are vacancies (in which an atom is missing); atoms out of place, occupying what is normally a hole in the lattice; substitutional impurities, in which foreign atoms occupy occasional sites; and interstitial impurities, in which foreign atoms occupy holes in the lattice.

Ionic (compound) lattices exhibit Schottky defects, in which a cation or anion is missing; and Frenkel defects, in which an ion leaves its site and enters a lattice hole. Line defects result from missing rows of atoms. Particularly common are edge defects, in which an extra partial plane of atoms is inserted near an edge of the crystal.

Example. Bending a paper clip. Take an ordinary paper clip and try to straighten it out. Are you able to get it perfectly straight in the regions where it was bent? Why not? Now try to rebend the paper clip to its original shape. Are you able to do it? Why not?

Solution. When the clip was manufactured, it was annealed following bending. Annealing is the process of heating the clip to soften it enough that atoms can diffuse into the gaps left in the bending process. In other words, annealing produces edge defects. When you attempt to straighten the clip, the edge defects prevent the alignment of atomic planes, so complete straightening is impossible. However, straightening the clip introduces new edge defects on what was originally the inside curve of the bend, so that when you attempt to rebend the clip, the new defects interfere.

Plane defects in crystals occur at the interface between two differently-oriented crystalline regions. As might be expected, such defects greatly diminish the resistance of the crystal to fracture. Crystal defects are important for a number of reasons. A problem with defects, especially of the line and plane type, is that they diminish the structural strength of the solid. Annealing is used to offset this.

Rusting of iron often occurs at defect sites on the surface, which may result when the metal is worked or penetrated by a rivet or bolt. However, defects can be advantageous as well. Defects are sites of high energy (*i.e.*, relative instability).

Reactions that take place at the surfaces of crystals often take place at these sites. This is involved in heterogeneous catalysis by solid materials, which is used on a huge industrial scale in our economy. Electronic components for modern computers require the reliable production of silicon that has a purity of 99.999999 per cent. Silicon crystals of such purity contain very few defects, which would diminish their functional effectiveness.

Silicon of this purity is produced by a process called zone refining. In this method, a bar of silicon of high purity is passed very slowly past a very narrow heating element, which melts a narrow region of the silicon bar. Impurities in the silicon dissolve in the melted part, and follow along with the melted region as the bar moves past the heating element. Eventually the fused region, containing all of the impurities, reaches the end of the bar, where it is solidified and cut off. The remaining silicon has the purity required in the manufacture of electronic components.

Properties of Liquids

We take up the liquid phase last because it is the most complex and least understood of the phases of matter. Liquids have properties intermediate between those of the solid and gas phases:

- A sample of liquid has a fixed volume (like a solid) and a molar volume only slightly greater than that of the solid. However,
- A liquid conforms to the shape of its container, like a gas, and
- A liquid flows readily, like a gas.
- Liquids are relatively incompressible. The relative compressibilities of the three phases are in the order g >> l > s. The incompressibility of liquids finds practical use in hydraulic machinery, such as automobile brakes.
- Diffusion is slower in liquids than in gases, but faster than in solids. The relative rates of diffusion in the three phases are in the order g >> l > s.

  These properties are readily observed visually. Several other properties of liquids are more subtle:

- Liquids give x-ray diffraction patterns that are diffuse, rather than sharp like those of solids. The pattern implies some order in the molecular arrangement, but less than in the solid.
- Liquids exhibit surface tension.
- Liquids exhibit viscosity.
- Liquids in closed containers evaporate to an extent. The resulting vapour exerts a pressure called the vapour pressure, P_{vap}, of the liquid. The magnitude of P_{vap} for a particular liquid depends only on temperature, not on the container size or the amount of liquid present. Liquid and vapour exist in equilibrium.

We begin with a molecular model for the liquid phase, based on its observable properties.

Model of the Liquid Phase

The small molar volume and incompressibility of liquids suggest that the molecules are close together. However, flow and the lack of a fixed shape indicate molecular mobility, despite the closeness. A feature of the model that is not evident from the figure is that the molecules are in constant motion, as in the gas and solid, colliding frequently with neighbours.

As for the gas and the solid phases, the kinetic energies of the particles of the liquid are described by the Maxwell-Boltzmann distribution, and the average KE per particle is 3/2 kT. Each molecule is fairly closely surrounded by other molecules, but there are definite gaps (holes) in the structure.

Molecules use these gaps to slip and slide easily past one another, which manifests macroscopically in the ability to flow. There are regions in the liquid that are quite ordered, similar to the solid, but the regions are constantly shifting position as molecules move to close gaps and open new ones. The gaps prevent the order from extending over long distances. Thus liquids have short-range order, in contrast to the long-range order of crystalline solids. The gaps allow diffusion of a molecule through the liquid, but frequent collisions with neighbouring molecules makes diffusion slow.

Contrast this with the rapid diffusion in the gas phase, where a molecule travels a long distance between collisions; and the extremely slow diffusion of the solid phase, where a molecule is locked into its lattice location. Diffusion rates in solid, liquid, and gas are best understood in terms of the mean free path, the average distance travelled by a molecule between collisions. The mean free path in the solid is virtually zero. In the liquid, it is a fraction of the molecular diameter. But in the gas, the mean free path may be 10-100 molecular diameters, depending on gas pressure.

Interesting Physical Properties Diffuse X-ray Diffraction Patterns

Indicates regions of short-range order in the liquid. Within these regions, molecules are almost close-packed. However, the gaps allow fluctuation in the distance of separation of the molecules. The nearly close-packed regions of the liquid diffract x-rays according to the Bragg equation, n = 2dsin*theta*. Fluctuations in d, a consequence of the gaps, cause fluctuations in *theta*, giving x-ray patterns that are diffuse, or smeared out.

This is a consequence of intermolecular forces. Consider a molecule at the surface of a liquid. In contrast to a molecule in the bulk, which is attracted equally from all directions, the surface molecule suffers an imbalance of forces because there are no molecules above it. The net force on it is directed towards the interior of the liquid. This pulling in of the liquid surface is called surface tension, symbolized z. The surface molecule has higher potential energy than a bulk molecule because of the lesser force on it.

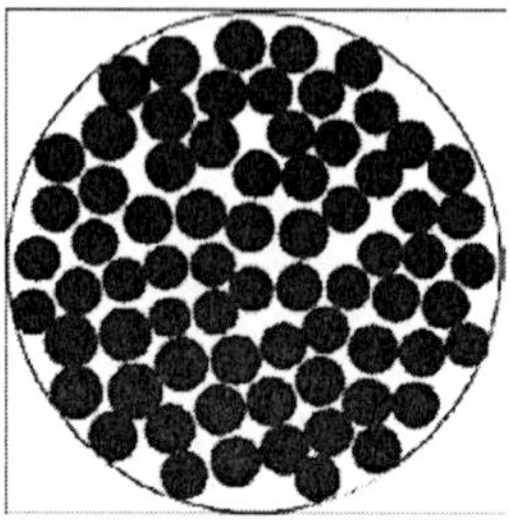

Surface Tension

To minimize energy, the liquid adopts a shape that minimizes the number of surface molecules. The shape with the smallest surface area-to-volume ratio is the sphere. Thus surface tension explains the spherical shape of raindrops and water droplets on a freshly waxed car: the spherical shape minimizes potential energy.

It is a safe prediction that manufacture of ball bearings will someday be carried out in space, where gravitational forces will not distort the sphericity that surface tension imposes on a drop of molten metal.

As temperature increases, molecular kinetic energy can overcome the attraction of intermolecular forces, and surface tension decreases according to equation.

$$z = a - bT$$

The constants a and b are characteristic of a particular liquid. For ethanol, $a = 24.05\times10^{-3}$ J/m^2 and $b = 0.0832\times10^{-3}$ J/m^2-°C. The units of surface tension are energy per area, consistent with the relation between potential energy and surface area discussed above.

A drop of water on a clean glass surface spreads and flattens, forming a film on the surface. This phenomenon of spreading and flattening is called wetting. Wetting occurs because forces between the surface molecules and molecules of the liquid exceed the intermolecular forces within the liquid. The liquid spreads to maximize the area of contact with the surface. Forces between unlike molecules are called adhesive forces; those between like molecules are called cohesive. Wetting occurs because adhesive forces exceed cohesive forces. If the reverse is true, a liquid will not wet a surface, but instead will sit on the surface in spherical droplets to minimize the area of contact with the surface. One visible consequence of wetting is the formation of a meniscus at the top of a column of liquid in a tube, such as a buret or pipet. The menisci formed by water and mercury.

The adhesive forces between the glass surface and water molecules are stronger than the cohesive forces between water molecules. Consequently, water achieves a maximum surface area of contact with the glass by bending up at the edges and sagging in the centre. The reverse is true for mercury, which minimizes the surface area of contact with the glass and forms a spherical surface shape to maximize cohesive forces.

The phenomenon of capillary action results from competition between surface tension and wetting. When one end of a capillary tube (a glass tube with inner diameter of approximately 1mm) is submerged in a liquid, a difference in height develops between the levels of liquid in the capillary and in the bulk liquid. A liquid that wets glass rises in the tube; a liquid that does not wet glass falls, called capillary action, for water and mercury. As water wets the inside wall of the capillary, the surface area of the water is increased.

However, surface tension tends to minimize the number of molecules at the surface, and causes the water to move up the tube, against the force of gravity, to offset the surface area increase resulting from wetting. The result is a capillary rise. For mercury, adhesive forces are weak, so the level in the capillary falls to minimize contact with the glass surface, and the surface of mercury in the capillary is pulled into a sphere by the strong cohesive forces. The result is capillary fall. Capillary action is important in the operation of blotters and sponges and is the mechanism by which water penetrates soils.

Viscosity

The viscosity of a liquid is a measure of its resistance to flow. A mobile liquid is one that flows readily (*e.g.*, water). A liquid that flows with difficulty (*e.g.*, molasses) is said to be viscous. Liquids composed of small, roughly spherical molecules are mobile, because the molecules can roll over one another smoothly, like ball bearings. Viscous liquids are composed of long, string-like molecules that easily become tangled during flow.

For example, the molecules of high viscosity motor oils consist of long chains of carbon atoms that intertwine and impede flow of the liquid. An extreme example of molecular tangling occurs in silly putty, a polymer that appears to be a solid but is in reality a liquid in which molecular tangling is so severe that hours are required for the putty to spontaneously change shape to that of the container.

Like surface tension, viscosity decreases (flow becomes easier) as temperature increases due to more rapid thermal motion of the molecules.

Vapour Pressure

Vapour pressure is the most fascinating physical property of liquids. It provides remarkable insight into the laws governing the behaviour of physical systems. We begin with a purely experimental approach to the phenomenon. The box is attached to a closed-end manometer and is connected via a stopcock to a vacuum pump.

A thermometer is inserted in the box and there is a septum-capped port (opening) that can be penetrated with a syringe needle. The stopcock to the vacuum pump is opened and the air is pumped out until the two arms of the manometer are at the same level. The stopcock is then closed. This is the initial situation, state 1, shown at the top left of the figure. Several experiments will now be performed in sequence.

Vapour pressure can be summarized as follows:

- A liquid in a closed container evaporates to some extent. The resulting vapour exerts a characteristic vapour pressure, P_{vap}.
- The value of P_{vap} is independent of the container size and the amount of liquid, provided that the amount exceeds a critical minimum.
- P_{vap} increases steeply with T.
- P_{vap} is a function only of the chemical identity of the liquid and the temperature.
- Liquids have a natural tendency to evaporate to some extent to form vapour.

The thought experiments, which can be verified readily in the laboratory, provide much descriptive information about the vapour pressure phenomenon. But the questions of why liquids tend to evaporate, and why they establish characteristic vapour pressures.

The Dissolving Power of Liquids

Liquids are invaluable to chemists as solvents. They are able to accommodate molecules of solute readily within the gaps in their structures to produce solutions that are themselves liquids. Such solutions are readily poured or otherwise transferred from one container to another; are conveniently measured out quantitatively using volumetric apparatus; and provide a uniform medium in which a chemical reaction can occur.

The unique ability of liquids to act as solvents is due to their dissolving power — the molecular motion of the liquid dislodges ions or molecules from the surface of crystals of a solid, then accommodates the separated ions or molecules in gaps. In this section we briefly discuss the dynamics of the dissolution process at the molecular level.

A microscopic view of two adjoining faces of a crystal of NaCl. A Na^+ ion on a face of the crystal differs from those in the interior because it experiences forces of attraction from only 5 nearest-neighbour Cl^- ions, rather than 6. Facial Cl^- ions are in a similar situation. Such ions are therefore in shallower potential wells than ions in the interior of the crystal and should be more easily dislodged.

Edge ions, such as the Na^+ ion labelled "a" in the figure, are attracted by only 4 nearest neighbours, and should be even more readily dislodged than face ions. Water molecules from the solvent continually bump against the faces and edges of the solid crystal. As they approach the crystal, their dipoles become oriented due to interactions with the face ion that they approach. Water molecules approaching cations have the O atom oriented towards the crystal face; those approaching anions put a H atom forward.

When a rapidly moving water molecule strikes a face or edge ion, sufficient kinetic energy can be transferred to dislodge the ion, which moves away from the crystal into a gap in the solvent structure (*e.g.*, Na^+ ion "b" in the figure).

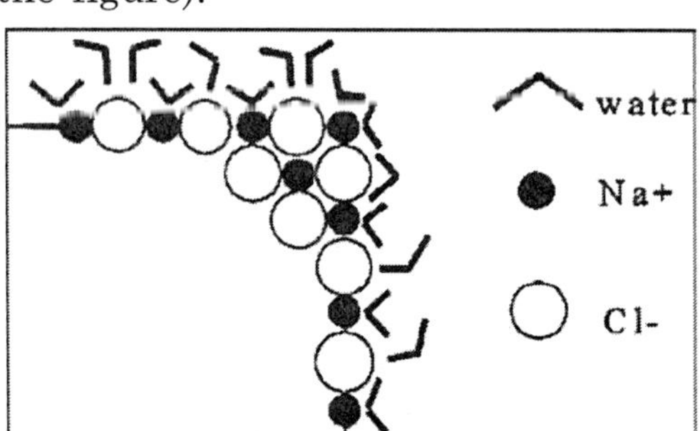

It becomes immediately surrounded by a cage of water molecules, with oxygen atoms nearest the Na^+ ion to maximize ion-dipole forces. The solvent cage insulates it from the crystal and makes return to the crystal face unlikely. Since edge ions are more readily dislodged than facial ions, and because they can be approached by solvent from several directions, dissolution occurs more rapidly at crystal edges than faces. The initially sharp edges of a dissolving crystal become rounded as dissolution procees.

Phase Transformations

A phase transformation is a conversion of a pure substance from on phase to another.

There are 6 types, each involving two of the three phases:

1. *Melting (fusion)*: Conversion of solid to liquid.
2. *Freezing*: Conversion of liquid to solid; the reverse of melting.
3. *Vaporization*: Conversion of liquid to vapour (gas). Its reverse is
4. *Condensation*: Conversion of vapour to liquid.
5. *Sublimation*: Cnversion of solid to vapour. Itsreverse is
6. *Deposition*: Conversion f vapour to solid.

The phase changes are illustratedin Figure. Since the solid and liquid phases have small molar volumes due to clse packing of molecules, they are referred to as condensed phases.

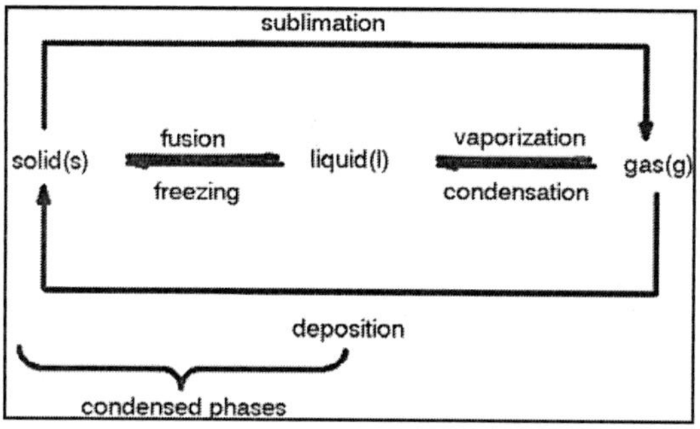

Figure: Phase Transformations

The two variables that determine the phase in which asubstance exists are the temperature and the pressure to which the substance is subjected. Phase transformations may therefore be accomplished in two limiting ways: by changing the temperature of the substance while keeping it under constant pressure; or by changing the pressure on the substance while keeping it at constant temperature. The first approach is the more familiar of the two. We make three assertions about temperature-induced phase changes:

- At a constant pressure, a phase transformation occurs at a constant temperature that is characteristic of the substance.

- To convert a given amount of substance from solid to liquid to gas requires addition of energy. This is usually accomplished by adding heat. The energies of the phases are therefore in the following order: $E_s < E_l < E_g$
- Conversion from solid to liquid o gas decreases the order (degree of organiztion) of the substance.

To clarify these assertions, we take a thought experimental approach to the phase changes of a substance, much as we did for vapour pressure. The results of these thought experiments can be readily verified in the laboratory. We start with a quantity of solid water (ice) in a syringe, state 1. The initial temperature of the ice and the inside of the syringe is -10°C. We plan to slowly add energy to the ice by controlled heating, which can be done by submerging the syringe in a large antifreeze bath with a hot plate under it. As we slowly add heat, we monitor the temperature of the water in the syringe with a thermometer inserted through an air-tight seal in the syringe. We begin at state 1.

Process: Turn on the hot plate and the bath stirrer. As the ice warms slowly from -10 to 0°C, nothing seems to happen. Temperature changes slowly and steadily. When the ice reaches 0°C, it begins to melt, and we see some liquid water form in the syringe.

Process: We continue to gently heat the bath. The ice gradually melts, but the temperature of the ice-water mixture remains constant at 0°C, despite the addition of heat. The temperature of the bath rises above 0°C, but as long as both ice and liquid water are in the syringe, T in the syringe remains at 0°. State 3 is reached when the last ice crystal melts. All water in the syringe is liquid, at T = 0°C.

Process: Once the ice has melted, the temperature of the water in the syringe begins to rise. As we slowly warm the bath, the water temperature slowly rises. Nothing happens in the syringe other than a slight increase in the volume of the liquid until T = 100°C. Then water begins to vaporize. This is the situation at state 4.

Process: We continue to slowly add heat to the bath and observe that temperature in the syringe remains constant at 100°C as long as both liquid and vapour are present, even though the bath temperature rises above 100°C. We also observe that as vapour forms, it pushes the plunger out. After all liquid has vaporized, at state 5, the temperature of the water in the syringe once more starts to rise.

Process: Continue heating the bath to T = 110°C. The temperature of the vapour in the syringe rises from 100° at state 5 to 110° at state 6, while its volume increases according to the ideal gas law. The pressure of the water vapour is maintained constant at 1 atm by movement of the syringe plunger.

These experiments verify the assertions at the beginning of the section. First, phase changes occur at constant T. This is shown by processes 2 and 4. The fixed, characteristic temperature at which a substance melts is called the melting point (s to l) or freezing point (l to s), and is symbolized T_f. The normal melting point is the temperature of melting under a pressure of 1.00 atm. For water, this is 0°C.

The characteristic temperature at which a substance boils under 1.0 atm pressure is called the normal boiling point. For water, this is 100°C. Second, process 2 shows that an input of energy is required to convert a solid to liquid. The same is true for conversion of liquid to gas (process 4). For each of these processes, the change in energy, DE ($E_{final}-E_{initial}$) is positive. Processes 2 and 4 may be represented as follows:

$$H_2O(s),\ 0°C \rightarrow H_2O(l),\ 0°C\ [DE > 0]$$

$$H_2O(l),\ 100°C \rightarrow H_2O(g),\ 100°C\ [DE > 0]$$

The process in equation occurs at constant T. Therefore the change in kinetic energy of the molecules is 0 (recall that an average molecule has KE of 3kT/2 at temperature T, regardless of phase). DE is therefore a change in potential energy, DPE. Thus the PE of liquid is higher than that of solid water at 0°C. Similarly, the PE of vapour is higher than that of liquid water at 100°C. Combining these conclusions gives equation:

$$PE_g > PE_l > PE_s \text{ at any T}$$

DPE > 0 for sublimation, fusion, and vaporization; DPE < 0 for the reverse processes. Third, the amount of disorder at the molecular level increases in our sequence of processes (we phrase this in terms of disorder rather than order). With addition of thermal energy (heat), the molecules convert from the long-range organized arrangement of the solid to the completely chaotic and random molecular distribution of the gas. Even warming the gas from 100 to 110°C decreases the order, because the volume available for the molecules to bounce around in increases. This conclusion is summarized in equation.

$$\text{disorder s} < \text{disorder l} < \text{disorder g}$$

Phases and the Potential Well, Revisited

The origin of the potential well in terms of electromagnetic forces within or between the fundamental units of substances. We concluded that forces are strongest and potential energy is minimized when fundamental particles form an orderly close-packed arrangement, and we recognized this as the situation in the solid phase. Heating the solid phase increases the kinetic energies of the particles of the solid phase until they are able to partially overcome the close-packed forces and separate somewhat. At this point, solid melts to liquid. Separation has increased the potential energy of the particles.

Further heating eventually gives the particles sufficient kinetic energy that they may escape the forces in the liquid into the space above the liquid; *i.e.*, vaporization occurs. In the vapour phase, molecules are so widely spaced that forces between them are negligible and potential energy has increased to zero. We thus see that phase conversion from solid to gas can be viewed in terms of particles climbing out of the potential well. Similarly, in the reverse conversion, particles fall into the potential well. The decrease in potential energy of the molecules is released as heat.

The Gas Phase

The phases (physical states) in which a pure substance may exist. For most substances there are three phases — solid (s), liquid (l), and gas (g). The gas phase is the simplest of the three to deal with theoretically, primarily because the molecules behave completely chaotically and therefore give rise to simply-formulated average properties. The solid phase is also relatively easily dealt with, because it is so highly organized.

Gases and their properties play an ubiquitous and critical role in our daily lives. Life is supported by the oxygen of the air, which reacts with glucose in biochemical combustion. Living systems have evolved complex molecular architecture to absorb oxygen from the air; carry it to the cells; deliver it to the cells; transport it within the cell; and supervise its multistep, controlled reaction with glucose. It is the rapid expansion of gases produced in the explosive combustion of gasoline that performs the work of the internal combustion engine. Expansion and contraction of gases in response to temperature and pressure changes is at the basis of continually changing weather patterns.

Clearly an understanding of the properties, behaviour, and chemical reactivity of gases is an important foundation for developing an appreciation of complex biological and technological systems. Following a discussion of the ideal gas law, a simple mathematical model for gas behaviour, we discuss the Kinetic Molecular Theory of gases, which establishes a link between molecular motion and temperature.

Macroscopic Properties of Gases

All gases have observable properties in common:

- Gases flow readily from one space to another and occupy all available space.
- Gases assume the shapes of their containers.
- Gases are readily compressible. By exerting force, we can cause a gas to occupy a smaller volume.
- Two or more gases form homogeneous mixtures (solutions) in all proportions. An example is air, a mixture of primarily nitrogen and oxygen with small amounts of other gases.
- Gases diffuse rapidly. This means that a gas can move across a large space in a relatively short time. Examples are the odors of perfume and skunk.

The state of a gas is described by the values of 4 macroscopic variables: the volume (V) occupied by the gas; the temperature (T) of the gas; the pressure (P) that the gas exerts on the walls of its container; and the amount of gas in moles (n). The values of these quantities are related by a very simple equation called the ideal gas law. We will discuss this equation after a brief word on pressure.

Pressure

Pressure is defined as force per unit area. We illustrate the concept of pressure with an example.

Example. The bottom of each foot of a 150-pound man has a surface area of about 24 square inches. When the man stands on a bathroom scale, what pressure is exerted by his feet?

Solution. The force exerted by the feet is the man's weight, 150 lb. This is distributed over the area of the bottoms of both feet, or 48 square inches. The pressure is then

$$P = F/A = 150 \text{ lb}/\ 48 \text{ in}^2 = 3.1 \text{ lb/in}^2$$

If the man stands on the scale with only one foot, the force is the same, but the pressure is doubled because the area is cut in half.

Gases exert pressure on the walls of their containers. Since a gas uniformly occupies a container, it exerts the same pressure on all the walls. The gases that make up the atmosphere exert pressure on the surface of the earth. This pressure, called the atmospheric pressure, can be measured using a barometer.

The left side of the figure shows a beaker or cup containing mercury. A glass tube of length 1 m and cross-sectional area, A, fitted with a stopcock on top, is inserted into the beaker so that the end of the tube is below the level of mercury in the beaker. Initially, the stopcock is open and the levels of mercury inside and outside the glass tube are the same. A vacuum pump is then attached to the top of the glass tube, and the air is pumped out. As this occurs, the mercury level rises in the tube until it reaches a height, h.

Despite further pumping, the mercury level rises no higher in the tube. When all air has been pumped out of the tube, the stopcock is closed, and the mercury maintains its position in the tube. This behaviour is understood as follows. Initially, before pumping, the atmosphere exerts the same pressure on the surfaces of mercury inside and outside the tube. Since $P_{out} = P_{in}$ = atmospheric pressure, the mercury levels inside and outside the tube are the same.

As air is pumped out of the tube, its pressure on the mercury inside the tube decreases, and mercury rises in the tube to a height, h, such that the force due to the mass of mercury in the tube just counterbalances the pressure of the atmosphere on the mercury outside the tube.

The height h is found to vary from day to day because the pressure of the atmosphere varies with weather conditions. However, its value is always in the neighbourhood of 0.76 meters. For this reason, a unit of pressure called the atmosphere (atm) is defined as the pressure that will support a column of mercury 76 cm, or 760 mm, in height.

$$1 \text{ atm} = 760 \text{ mm Hg}$$

The pressure exerted by a column of mercury 1 mm in height is called a torr after Evangelista Torricelli, inventor of the barometer. Thus

$$1 \text{ atm} = 760 \text{ mm Hg} = 760 \text{ torr}$$

In the English and SI systems, the units of pressure are respectively the pound per square inch and the pascal (1 Pascal = 1 Newton per square meter, where a Newton is the force required to accelerate a 1 kg body at 1 m/s^2). The pressure in Pa exerted by a column of mercury 760 mm in height can be readily calculated and will provide a conversion factor between SI and traditional non-SI units. The force exerted by a mass, m, of mercury is mg, where g is the acceleration due to gravity. This force is distributed over the cross sectionl area, A, of the tube. The mercury therefore exerts the pressure, P = mg/ A on the surface of mercury at the bottom of the tube. The mass of mercury can be calculated from its density, r, and its volume, which is Ah. Substituting for m in the expression for pressure gives equation:

$$P = \rho \times (Ah)g/A = \rho \times gh$$

The pressure exerted by a column of mercury (or any liquid) of height h is independent of the cross sectional area of the column. The diameter of the barometer tube is therefore unimportant. Substituting the density of mercury (13.595 g/cm^3), the acceleration constant (g = 9.80665 m/s^2), and h = 0.760 m in 5–1–1 gives $P = 1.0132 \times 10^5\ Nm^{-2}$ ($1\ Nm^{-2}$ = 1 Pascal). The following conversion relations are thus obtained:

$$1\ atm = 14.7\ lb/in^2 = 1.013 \times 10^5\ Pa\ (1\ Pa = 1 N/m^2)$$
$$= 1.013\ bar\ (1\ bar = 10^5\ Pa)$$

Although the bar is official, the atmosphere and the torr are still the most commonly used pressure unitsin chemistry. We will therefore use the atm and the torr. Note that atmospheric pressure is 14.7/3.1 = 4.7 times greater than the pressure exerted by the feet of the man in Example. The magnitude of the pressure exerted by the atmosphere is equivalent to having that 150-lb man, holding a 550-lb weight, standing on your stomach.

The pressure of a gas confined in a container cannot be measured with a barometer. For this a variety of devices may be used, the simplest being the manometer. A manometer is a U-shaped glass tube containing mercury or some less toxic fluid. The gas container is attached to one side of the manometer, and the other side is either open to the atmosphere (an open-end manometer) or evacuated (a closed-end manometer). The difference in the levels of fluid in the two arms of the manometer is a measure of the pressure of gas in the bulb through the relationship, $P_1 = P_2 + P$. Here P_1 is the pressure exerted by the gas on arm 1 of the manometer, P_2 is the pressure exerted on the arm of the manometer not connected to the gas container (arm 2), and P is the height of fluid in arm 2 minus the height in arm 1. P is positive if fluid is higher in arm 2 and negative if it is higher in arm 1.

For an open-end manometer, P_2 is atmospheric pressure; for a closed-end manometer, P_2 is zero. If the manometer fluid is mercury, the pressure is in torr and may be converted to atm using the conversion factor given above. If the manometer contains some other fluid, the height of fluid may be converted to an equivalent height of mercury using the densities of the fluid and mercury.

Example. An open-end manometer containing water as the working fluid is used to measure the pressure of gas in a bulb. The level of water in the side attached to the bulb is 8.65 cm higher than that in the side open to the atmosphere. Atmospheric pressure is 693 torr. What is the pressure of gas in the bulb?

Solution. Use the relation P = r×g×h. To find the height of mercury equivalent to 8.65 cm of water, we equate r×gh for water to r×gh for mercury. The gravitational acceleration cancels, giving

$$(1.00 \text{ g/cm}^3)(86.5 \text{ mm}) = (13.60 \text{ g/cm}^3)(\text{h})\text{h(Hg)} = 6.4 \text{ mm}$$

The pressure of gas is this amount less than atmospheric pressure:

$$P_{gas} = 693 - 6.4 = 687 \text{ torr.}$$

The Ideal Gas Law

The ideal gas law is a relationship between the pressure (P), the volume (V), the amount (n), and the temperature (T) of a gas. The ideal gas law is called an equation of state, because it represents all possible states — combinations of P,V and T — in which the amount of gas, n, can be found.

$$PV = nRT$$

In this equation, P is the pressure in atm; V is the volume in L; n is the amount of gas in moles, and T is the Kelvin temperature. R is a proportionality constant, remarkably the same for all gases, which relates the 4 variables. Equation is of simple form, which belies the effort expended in developing it. Over 200 years elapsed between the first quantitative experiments on gases and the final enunciation of equation. The simple form of the equation hides a remarkable sophistication in understanding. We now examine what the equation says regarding gas behaviour.

Volume-Pressure Relation

If we rearrange the equation to solve for the volume, the equation states that volume decreases as the pressure exerted on (also by) the gas increases. The inverse relation of pressure and volume is a fact of every day experience. For example, when you compress the air in the chamber of a tire pump, you feel the increase in pressure of the air within.

In 1662, Robert Boyle developed the quantitative relationship between pressure and volume that we now know as Boyle's Law, which states that the volume occupied by a fixed amount of gas at a fixed temperature varies inversely with the pressure exerted on the gas. Boyle's Law was the first-discovered component of equation. Note that the ideal gas law predicts that the volume should fall to 0 as pressure gets very large. For real gases this does not happen.

At ordinary temperatures, most real gases will liquefy if pressure is made large enough. Once liquefication has occured, volume does not decrease much with increasing pressure. Real gases obey equation closely only when P is not too large. This is why equation is called the ideal gas law — it applies strictly only to an idealized gas, which does not liquefy but instead shrinks to zero volume when pressure is made very large.

Volume-Temperature Relation

The ideal gas law states that the volume occupied by an amount of gas, n, varies directly with the absolute temperature of the gas. A particular sample of gas, then, should double in volume if heated from 300 to 600 K. This behaviour, too, is quite well known, at least qualitatively, in every day experience, as in the expansion (contraction) of a heated (cooled) balloon.

The relationship between gas volume and Celsius temperature was first established by Jacques Charles in 1800. He clearly established the linear relationshipbetween the two quantities. Similar experiments relating gas pressure and temperature were carried out at about the same time by Gay-Lussac, and similar proportionality was found. In both cases, the relationships obtained were not so simple as the ideal gas law, because the absolute scale of temperature was not then recognized. The relation between volume and Celsius temperature is in equation.

$$V = a \times t(^{\circ}C) + b$$

Plots of V versus t(°C) for gas samples containing different amounts of gas give different slopes and different intercepts on the V axis. However, they all give the same intercept on the t axis, -273.16 °C. This suggests that the zero of temperature be shifted 273.16 Celsus degrees to the left. The resulting scale of temperature is, of course, the Kelvin scale. Equation takes the much simpler form, V = a×T.

A profound implication of equation is that there exists a temperature at which the volume of gas should become zero. This temperature is called absolute zero. The Kelvin scale is set up so that 0 K corresponds to the temperature at which the volume of an ideal gas sample would become zero. This corresponds to -273.16 on the Celsius scale. Any temperature lower than this value would cause the gas to have negative volume. Since this is physically meaningless, the conclusion is that temperatures lower than absolute zero do not exist.

That there is a lower limit on attainable temperature is not obvious from our everyday experience and from the manner in which we normally measure and think about temperature; the conclusion is nonetheless correct.

Relation Between Volume and Amount of Gas

The ideal gas law states that the volume occupied should increase in direct proportion with the amount (moles) of gas, when the gas is kept at the same pressure and temperature throughout. This is an extension of the hypothesis by Amadeo Avogadro in 1811 that equal volumes of different gases contain the same number of molecules.

Avogadro made this statement in explanation of many observations about the relative volumes of gaseous elements that react to form compounds. It was in fact this hypothesis that finally allowed correct molecular formulas to be deduced and the atomic mass scale to be put on a firm basis.

The Value of the Gas Constant R

The gas constant R turns out to be an ubiquitous and extremely important quantity in physcal science. We begin by obtaining a value for it based on the experimentally-measured volume of 1.0 mole of gas at 1.00 atm pressure and a temperature of 273.16 K. It is found that under these conditions the gas occupies a volume of 22.414 L. Rearranging equation to solve for R, we can calculate its value:

R = PV/nT = (1.000 atm)(22.414 L)/ (1.000mole) (273.16K)
= 0.08206 L-atm/K-mole

The numerical value of R depends upon the units in which pressure and volume are expressed. P is commonly expressed in atmospheres or torr (both non-SI) or Pascal (SI), and V is commonly expressed in mL. Values of R in common P-V unit combinations are presented in Table.

Table: In Common Pressure-Volume Unit Systems

Pressure Unit	*Volume Unit*	*R*
Atm	L	0.08206 L-atm/K-mole
Bar	L	0.08314 L – bar/K-mole
Torr	mL	6.237×10^4mL-torr/K-mole
Pa	m^3	8.314 J/K-mole

From the Pascal-m^3 value, we see that R has units of energy/K-mole. It follows that the combined units L-bar, L-atm, and mL-torr are also energy units. We can express R in terms of any desired energy unit by using an appropriate conversion factor. In using the ideal gas law, equation, the units of P, V, and R must be self-consistent, and the temperature must be expressed in Kelvins. A few examples of the use of the ideal gas law are now presented.

Example. An ideal gas occupies 326 mL at 730 torr of pressure and a temperature of 19°C. What is its volume at 1.00 atm and 0°C?

Solution. We take two approaches to the solution of this problem. In the first approach, we use the ideal gas law twice. First, we calculate the moles of gas under the original V,T,P conditions; second, we calculate the volume occupied by this number of moles at 1 atm and 0°C.Convert to the necessary units of T, P:

Initial	*Final*
T(K) = 273 + 19 = 292 K	T(K) = 273 K
P = 730 torr = 730/760 atm = 0.961 atm	P = 1.00 atm
V = 326 mL = 0.326 L	V is unknown

Calculate the number of moles of gas under the initial conditions:

$$n = PV/RT$$

$$= (0.961 \text{ atm})(0.326 \text{ L})/(0.08206 \text{ L-atm/K-mole})(292\text{K})$$

$$= 1.307\times10^{-2} \text{ moles}$$

Calculate volume at final T and P:

$$V = nRT/P$$

$$= (1.307\times10^{-2} \text{ moles})(0.08206$$

$$\text{L-atm/K-mole})(273\text{K})/ (1.00 \text{ atm})$$

= 0.29 L (2 significant figures)

To shortcut this long process we take advantage of the constancy of the amount (moles) of gas. Since n is constant, PV/RT must also be constant, and its initial and final values are the same:

$$P_iV_i/T_i = P_fV_f/T_f$$

Solving for V_f and substituting numbers gives $V_f = 0.29$ L.

Example. 3.24×10^{-4} moles of ideal gas occupies 5.31 mL at 25°C. Calculate the gas pressure.

Solution. Convert units, and apply the ideal gas law to the problem.

Convert units:T(K) = t(°C) + 273 = 25 + 273 = 298KV(L) = 5.31 mL/(1000 mL/L) = 5.31×10^{-3} LCalculate P:P = nRT/V = $(3.24\times10^{-4})(0.08206)(298)/(0.00531) = 1.49$ atm

Example. Consider the apparatus in Figure. Initially, a quantity of gas is confined in the left box at a pressure of 3.40 atm and a temperature of 286 K. 4 sequential processes, a, b, c, and d, are carried out. Work out the value of each indicated quantity after the completion of each process.

- Valve 1 is opened with valve 2 closed. T is kept constant. What are the final volume and pressure of the gas?
- T is raised to 350 K. What are the final V and P of the gas?
- Valve 1 is closed, valve 2 is opened. What are the final V and P of the gas to the right of valve 1?
- T is lowered to 300 K. What are the final V and P of the gas to the right of valve 1?

Solution. We apply the ideal gas law to each process in turn.

- When valve 1 is opened, the gas will expand to occupy its original box and the 3.0-L box at uniform pressure. T is constant, and the amount of gas is too. The gas volume goes up by a factor of 4, so the pressure goes down by the same factor.

  V occupied by gas = 4.0 LP of gas = 3.4/4 = 0.85 atm
- When T is raised to 350 K, the gas pressure will increase in direct proportion, because V will remain constant. Therefore

  V occupied by gas = 4.0 LP of gas = T_fP_i/T_i = (350)(0.85)/(286) = 1.04 atm
- When valve 1 is closed, the gas in the 1.0-L box will no longer communicate with the gas in the middle box. Only the gas in

the middle box will be involved when valve 2 is opened. This gas will expand from its initial pressure of 1.04 atm until its final pressure is 0.25 atm. At this point, the piston in the cylinder will stop moving.

V occupied by gas to right of valve 1 = P_iV_i/P_f = (1.04)(3.0)/(0.25) = 12.48 L.P of gas to the right of valve 1 = 0.25 atm

- When T is lowered to 300 K, the volume occupied by the gas will decrease, as the piston moves in to maintain pressure at 0.25 atm.

  V occupied by gas to the right of valve 1 = V_iT_f/T_i = (12.48 L)(300)/(350) = 10.70 L.P of gas to the right of valve 1 = 0.25 atm

Example. 0.432 g of an ideal gas exerts pressure of 0.46 atm in a volume of 768 mL at 0°C. Calculate the molar mass of the gas.

Solution. The ideal gas law can be expressed directly in terms of MM because moles is mass over MM:

n = PV/RT = (mass)/MM

Rearrange for MM:

MM = mRT/PV, where m is mass.

Solve:

MM =(0.432g)(0.08206 L-atm/K-mole)(273 K)/(0.46 atm)(0.768 L)
= 27.4 g/mole

The gas could be nitrogen, N_2; carbon monoxide, CO; or ethylene, C_2H_4.

This method for determining the MM of an unknown gas or volatile liquid is called the Dumas method. It is simple, fast, and reasonable accurate.

Example. Calculate the density of oxygen gas at a pressure of 0.940 atm at room temperature (25°C).

Solution. Rearrangement of the ideal gas law with the recognition that density is mass per volume gives density directly.

Rearrange:

PV = nRT

n/V = m/(MM)V = r/MM = P/RT

ρ = P(MM)/RT

Solve the equation with the specific variables given:

= (0.940 atm)(32.00 g/mole)/(0.08206 L-atm/K-mole)(298K)

= 1.23 g/L

The value has units g/L, and is about 1000 times less than the density of water at the same temperature. Gases are much less dense than liquids, implying that the gas molecules are much further apart than are liquid molecules.

Dalton's Law of Partial Pressure

The ideal gas law can be applied to mixtures of gases as well as to single pure gases. Consider a mixture of two non-reacting gases, A and B, in a container of volume V. According to the ideal gas law, the pressure exerted by the mixture is given by equation.

$$P_T = n_T RT/V$$

Here n_T, the total moles of gas, is the sum of the moles of A and the moles of B:

$$n_T = n_A + n_B$$

Substitute equation and expand:

$$P_T = (n_A + n_B)RT/V = n_A RT/V + n_B RT/V$$

Suppose that we could remove all of the molecules of B from the container. What pressure would A exert in the container alone? This is easily obtained from the ideal gas law to be $P_A = n_A RT/V$. Similarly, $P_B = n_B RT/V$. Substituting the expressions for P_A and P_B into equation gives:

$$P_T = P_A + P_B$$

This is Dalton's Law of partial pressures. In words, it says that the total pressure of a mixture of gases (P_T) is the sum of the partial pressures of the gases composing the mixture. The partial pressure of gas A is the pressure that n_A moles of A would exert if present alone in the container. Dividing the expression for P_A by equation gives:

$$P_A/P_T = n_A/n_T = n_A/(n_A + n_B)$$

The ratio of the moles of A to the total moles of all gases in the system is called the mole fraction of A, and is symbolized X_A. Thus $X_A = n_A/n_T$. Substituting this into 5-2-7 and rearranging, we obtain equation:

$$P_A = X_A P_T$$

The partial pressure of A in a mixture of A and other gases is the total pressure multiplied by the mole fraction of A.

Example. A gas mixture containing equal numbers of nitrogen and oxygen molecules exerts 1.32 atm pressure. What is the partial pressure of nitrogen?

Solution. Since half the molecules are nitrogen, half the moles are also nitrogen, so the mole fraction of nitrogen is 0.5.

$$P_{N2} = 0.5\ (1.32\ \text{atm}) = 0.66\ \text{atm}$$

We close this section by emphasizing an implicit assumption of the above treatment. Figure shows two gases, A and B, in a box.

If the box has volume V, what fraction of this volume is occupied by A, and what fraction by B? A frequent answer might be that since half the molecules in the box are A, A occupies half the volume. This is, however, incorrect. If, when we fill the box with gas, we put A in first, the molecules of A will spread out to occupy the whole box. Now add some molecules of B. Since these molecules are "unaware" of the presence of A, they, too, will spread out in the entire volume. Thus both gases occupy the same volume — the entire volume of the container. $V_A = V_B = V$, the box volume. This is implicit in equations. Similarly, we assumed that temperature is the same for both gases: $T_A = T_B = T$. In contrast, the total moles, n_T, is the sum of the moles of A and B; and the total pressure is the sum of the partial pressures. These relationships are summarized in Table.

Table: Relationships Among State Variables for a Gas Mixture

$$T_A = T_B = T_{mixture}$$
$$n_A + n_B = n_{mixture}$$
$$P_A + P_B = P_{mixture}$$
$$V_A = V_B = V_{mixture}$$

Dalton's Law is particularly useful in experimental situations involving the collection of gases over liquids. In such situations, the space above the liquid contains not only molecules of the gas, but also molecules of the vapour form of the liquid, produced by evaporation of the liquid. The vapour exerts a partial pressure that is part of the total pressure exerted by the gas mixture over the liquid.

This partial pressure du to vapour that exists above the corresponding liquid phase is the vapour pressure, P_{vap}, of the liquid. According to Dalton's Law,

$$P_T = P_{gas} + (P_{vap})\text{liquid}$$

To calculate P_{gas}, we must correct the total pressure for the vapour pressure of the liquid, which we can look up in a table.

Example: 0.1140 g of an unknown metal M reacts with excess aqueous HCl according to the reaction

$$M + 2H^+ \rightarrow M^{2+} + H_2(g).$$

The $H_2(g)$ is collected over water. The volume of the gas above the water is 52.1 mL, the total pressure of the gas is 738 torr, and the temperature is 22.1°C. The vapour pressure of water at 22.1°C is 20.0 torr. Calculate the molar mass of the metal.

Solution: Strategy: $P_T \rightarrow P_{H2} \rightarrow$ moles $H_2 \rightarrow$ moles M $\rightarrow$ molar mass M.Use Dalton's Law to calculate the pressure of H_2 in the gas mixture above the water:

$$P_{total} = P_{H2} + (P_{vap})H_2O$$

$$P_{H2} = P_{total} - (P_{vap})$$

$$H_2O = 738–20 = 718 \text{ torr}$$

Calculate the moles of H_2 from the ideal gas law:

$$n_{H2} = PV/RT = (718/760 \text{ atm})(0.0521 \text{ L})/(0.08206 \text{ L-atm/K-mole})(295.1 \text{ K})$$

$$= 2.033\times10^{-3} \text{ moles}$$

Calculate moles M from stoichiometry:

moles M = 2.033×10^{-3} moles $H_2\times$1mole M/1 mole H_2 = 2.033×10^{-3} moles M

Calculate molar mass of M from known mass and moles:

MM = 0.1140g/2.033×10^{-3} moles = 56.1 g/mole

The metal is probably Fe.

Example. Figure shows a syringe that has been filled with $N_2(g)$ to a volume of 20.0 mL. The end of the syringe has been capped with a rubber septum (a membrane that may be penetrated to add or remove substances from the syringe, but which otherwise seals the syringe off). 2.50 mL of pure liquid ethanol, $C_2H_6O(l)$, is injected into the syringe. It is observed that the syringe piston moves out until the occupied volume in the syringe is 32.00 mL. Atmospheric pressure is 730.0 torr, and the temperature of the syringe and its contents is 25.00°C. Calculate the pressure exerted by ethanol vapour in the syringe.Solution. See if you can figure this out on your own.

Kinetic Molecular Theory

The ideal gas law was developed from macroscopic observations, with no knowledge of the behaviour of a gas at the molecular level. We now attempt to interpret the ideal gas law in terms of a reasonable molecular model (simple physical picture) of gas behaviour.

What are the individual molecules doing, and why does their behaviour manifest itself in the ideal gas law? We now enter the domain of theory — a mental interpretation of experimental results. The model of gas behaviour currently accepted by scientists is the Kinetic Molecular model. The theory of gas behaviour built on this model is called the Kinetic Molecular Theory (KMT). This theory is one of the oldest and most resoundingly sccessful in science.

First, we assume that gases consist of very small molecules. Since a gas occupies the entire volume of its container, and since gases readily diffuse, the molecules must move about in space. Consequently they have kinetic energy — energy of motion. Since gases are compressible, the molecules must be far apart. Focussing on an individual molecule, it seems reasonable to assume that it is a small particle that moves about at random, occasionally colliding with a wall of the container and with other gas molecules.

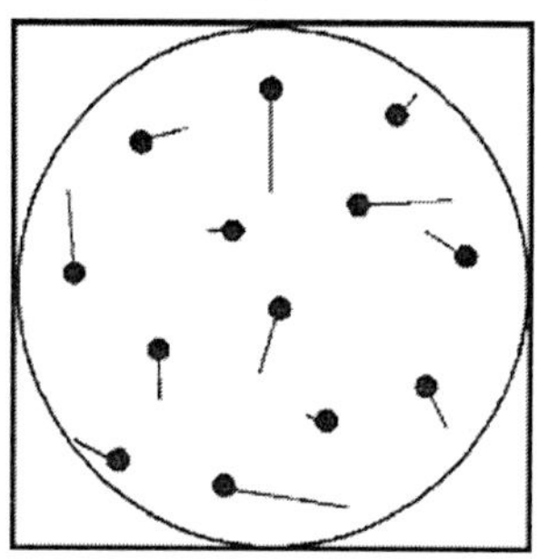

Figure: Dynamic Model of the Gas Phase

The molecule sometimes gains and sometimes loses energy in these collisions, so that it frequently changes speed. Sometimes it moves rapidly, sometimes slowly. Thus our picture of the gas is dynamic rather than static. Figure is an attempt to portray this simple physical model of the gas phase.

These statements about gas molecules are consistent with the properties of gases listed. None of them have been proven, but they seem reasonable and constitute the postulates of the Kinetic Molecular Theory:

- Gases consist of molecules that behave like tiny hard spheres;
- The molecules have no volume — they may be treated as points;
- There are no forces of attraction between molecules;
- Collisions between molecules are elastic (kinetic energy is conserved);
- The molecules are in ceaseless random motion, frequently colliding with the container walls and with each other. A collection of gas molecules is highly disordered.
- The average kinetic energy of the molecules is proportional to Kelvin temperature.

We now pursue the consequences of these postulates. Postulate 6, the origin of which is not obvious. Our aim is to develop an expression for pressure. At the molecular level, pressure must result from collisions of gas molecules with the container walls. A molecule hitting the wall exerts a force on it. The collection of all such forces on a unit area of wall during a given time interval constitutes the pressure. We should be able to calculate P from the force exerted by each collision, multiplied by the number of collisions per unit area of wall:

P = Force/Area = Force/collision×Collisions/area

From Newton's second law, force is the rate of change of momentum. Thus

F = momentum change/time-collision×collisions/area

We move the time factor over to the second term to obtain equation.

F = momentum change/collision×collisions/time-area

To obtain the momentum change per collision, envision a molecule moving directly towards a wall with velocity v, as shown in figure.

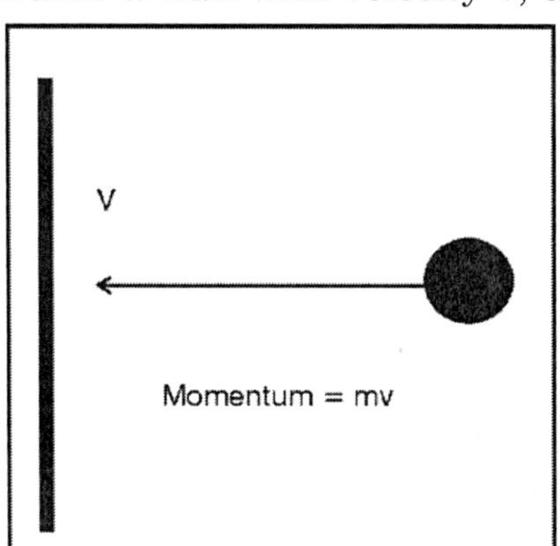

Figure: Collision of Gas Molecule with Container Wall

Since momentum is conserved in any collision, the change in momentum of the molecule is momentum change = momentum after collision–momentum before= m(-v)–m(v) = -2mv The change of momentum of the wall must therefore be 2mv to give a total change of zero. We now have the first term in equation:

$$\text{momentum change/collision} = 2mv$$

We take an intuitive approach to the second term. The number of collisions per unit time per unit area of wall should depend on 1) the number of molecules per unit volume in the container, N/V (the more there are, the more collisions there should be with the walls); and 2) the speed v at which a molecule moves (the faster the movement, the more collisions per unit time that should occur). Equation is based on these ideas:

$$\text{collisions/time-area} = (N/V)(v)$$

If we assume for simplicity that the container is cubical, there are 6 walls over which these collisions must be spread. The collisions with a particular wall are then

$$\text{collisions/time-area} = Nv/6V$$

(Despite our restrictive assumptions about the shape of the container, equation turns out to be valid for a container of any shape!) We now multiply the expressions in equations to obtain the pressure:

$$P = (2mv)(Nv/6V) = Nmv^2/3V$$

Rearrangment gives equation.

$$PV = Nmv^2/3$$

This is as far as our molecular model takes us.

We are now at the interface between theory and experiment. Experiment (the ideal gas law) relates pressure and volume to temperature:

$$PV = nRT$$

Theory, equation, relates pressure and volume to the mass and speed of a gas molecule. To bring experiment and theory into correspondence, we must equate the right sides of the preceding two equations. In order for KMT to successfully explain the ideal gas law, it is necessary that

$$nRT = Nmv^2/3$$

Equation can be simplified by replacing n with N/N_0 (N_0 is Avogadro's Number); $mv^2/2$ by KE (the kinetic energy of an average gas molecule); and solving for KE. The result is

$$KE_{molecule} = 3RT/2N_o$$

We refine this equation by making two further realizations. First, since R and N_o are both constants of nature, so must their ratio be. It is symbolized k, and is called Boltzmann's constant. Its value is 1.381×10^{-23} J/K-particle. Second, we have recognized that the kinetic energy of a gas molecule changes frequently as it undergoes collisions. However, its average speed over time is constant. The refined equation is among the most importantequations in science:

$$\text{Average } KE_{molecule} = 3kT/2$$

The average kinetic energy of a gas molecule, no matter what its chemical identity, depends only on the Kelvin temperature. This gives a deep insight into the concept of temperature: it is a measure of the average kinetic energies of the molecules of a substance, hence a measure of their average speed. As T is increased, molecules move faster; as it is lowered, they move slower. The temperature at which molecular motion ceases is absolute zero. Temperatures lower than absolute zero are impossible because a molecule may not have negative kinetic energy. Equation is simple and profound. Is there an experiment that can be done to test its validity? There is in fact a simple experiment that verifies equation in a rearranged form. If we replace the average kinetic energy of a molecule with the expression $m(v^2)_{avg}/2$, where $(v^2)_{avg}$ is the average of the square of the speed, and solve the resulting expression for $(v^2)_{avg}$, we obtain equation.

$$(v^2)_{avg} = 3kT/m$$

The square root of $(v^2)_{avg}$ is called the root mean square speed, and is for our purposes approximately equal to the average molecular speed. Making this equality gives equation.

$$v_{avg} = (3kT/m)^{1/2}$$

Multiplying both numerator and denominator of the argument on the right side of this equation by Avogadro's number N_o gives the useful variant in equation.

$$v_{avg} = (3RT/MM)^{1/2}$$

The implication of equations is that a heavy gas molecule moves more slowly than a light one, in a quantifiable way. The ratio of the speeds of the light (L) and heavy (H) molecules should be the square root of the inverse ratio of their molar masses:

$$(v_H/v_L)_{avg} = (MM_L/MM_H)^{1/2}$$

In 1846, Thomas Graham measured the rates of diffusion of various gases. Diffusion is the process by which gases disperse in space via random molecular motion. The results of Graham's experiments are summarized in equation rate diffusion of gas A/rate of diffusion of gas B =(density of gas B/density of gas A)$^{1/2}$ But we have previously seen that gas density is proportional to MM. If we make the logical assumption that a gas diffuses at a rate that is directly proportional to the average speed of its molecules, then Graham's Law of Diffusion is in exact agreement with equation.

Example: Calculate the ratio of the speeds of H_2 and CO_2 molecules at 25°C.

Solution: As long as temperature is the same for both molecules, its actual value is unimportant.

$$v_{H2}/v_{CO2} = (MM_{CO2}/MM_{H2})^{1/2} = (44.0/2.02)^{1/2} = 4.67$$

Example. Calculate the average speed of a molecule of nitrogen gas at room temperature, 25°C.

Solution. $v_{N2} = (3RT/MM)^{1/2} = (3(8.314\text{ J/K-mole})(298\text{ K})/(0.02801\text{ kg/mole}))^{1/2} = 5.15\times10^2$ m/s

This is 1150 miles per hour, or somewhere between the speeds of a commercial jet liner and a fighter jet.

Please note that in applying equation in Example, values of R and MM with appropriate units have been used, so that speed comes out with appropriate units. Use of R in units of, for example, atm-L, produces a speed with absurd units.

The curve is a plot of the fraction of molecules having speed s, f(s) (we use s for speed, v for velocity), versus the possible values of speed between zero and infinity. The curve has several noticeable features:

- The curve has the expected shape — low on both ends, indicating that few molecules have extremes of speed; and peaking in the middle, indicating that most molecules have speeds somewhere between the extremes.
- The curve s unsymmetrical because there is a definite lower limit (zero) but no definite upper limit to speed.
- The total area under the curve is the total number of gas molecules.
- The height of the curve at a particular speed is proportional to the fraction of molecules having speed s_1.

- The area under the curve between two speeds s_1 and s_2 (shaded in the figure) is the fraction of molecules haveing speeds in the range s_1 to s_2.
- The mathematical expression for f(s) obtained by Maxwell and Boltzmann is equation:

$$f(s) = (\text{constant})(s^2)\ \exp(-ms^2/2kT)$$

The s^2 term increases with increasing s and the exponential term decreases with increasing s. Thus f(s) first increases, but then peaks and decreases with increasing s as the exponential term takes over. The speed at the maximum of the curve, s_{mp}, is the most probable speed, because the greatest fraction of molecules have it.

The Maxwell-Boltzmann Distribution Law. We have said that in a gas the molecules move randomly, and that the speed of a molecule changes frequently. In the mid 1800's,James Clerk Maxwell n England, and Ludwig Boltzmann in Austria, were concerned with using statistical methods to describe, precisely and mathematically, the distribution of molecular speeds. The result of their efforts is called the Maxwell-Boltzmann istribution Law. They showed that at a given temperature, the molecular speed distribution follows a curve like that in Figure.

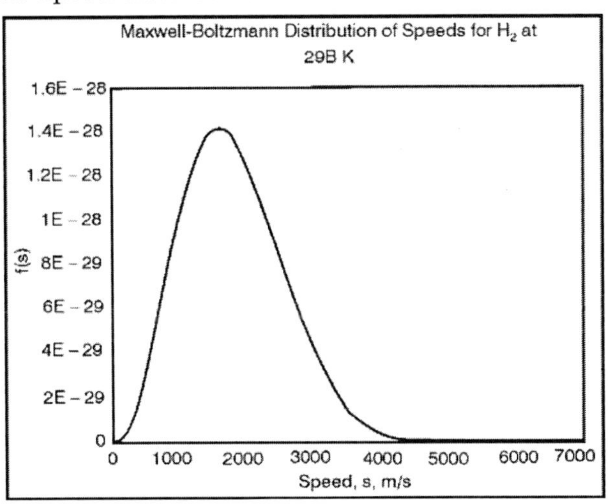

Figure: Maxwell-Boltzmann Distribution

- As T increases, the curve flattens and the maximum moves to higher speed, because the speeds of all molecules tend to increase with increasing T. The area under the curve stays constant, however, as long as the number of gas molecules is unchanged.

We now do two calculations using equation. First, we calculate the most probable speed, s_{mp}. This is a good approximation to the average speed, but the average will be somewhat larger because the curve is biased to higher speeds. s_{mp} is the speed at which f(s) is maximum. At this maximum, the slope of the plot — the derivative of f(s) — is zero:

$$df(s)/ds = d/ds\ [(s^2)(\exp(-ms^2/2kT))] = 0$$
$$= ds^2/ds\ \exp(-ms^2/2kT) + s^2\ d[\exp(-ms^2/2kT)]/ds = 0$$
$$= 2s\ \exp(-ms^2/2kT) - s^2(ms/kT)\exp(-ms^2/2kT) = 0$$

Dividing by the exponential, we obtain

$$2s - ms^3/kT = 0$$

This is readily solved for s_{mp}:

$$s_{mp} = (2kT/m)^{1/2}$$

As expected, this is a bit smaller than the root-mean-square speed in equation.

Next we calculate the average KE of a gas molecule by a method of averaging; we multiply each possible kinetic energy by the number of molecules with that kinetic energy, and divide by the total number of molecules. The required mathematical procedure is integration.

$$KE_{molecule} = \text{Integral from 0 to Infinity of } f(s)(ms^2/2)ds = 3kT/2$$

Even if you are not yet able to carry out the integration, you should note that the result is the same as equation, which came from KMT. This is gratifying. It reinforces the validities of both KMT and the M-B Distribution Law.

Finally, we calculate the KE per mole of gas from the average $KE_{molecule}$:

$$KE_{mole} = N_o \times KE_{molecule} = 3RT/2$$

The kinetic energy of a mole of gas depends only on temperature. For this reason, kinetic energy of molecules is called thermal energy. Summary of the Interpretation of the Ideal Gas Law in Terms of Molecular Behaviour. The relationship between the macroscopic and microscopic views of gas behaviour can be summarized in several statements.

- Gases are compressible because they consist of small molecules that are far apart.

- Molecules of a gas are in constant motion, allowing gases to diffuse and to flow.
- Pressure results from collisions of gas molecules with the container walls.
- Pressure increases as volume decreases (Boyle's Law). A decrease in V increases the number of molecules per unit volume, leading to more molecular collisions with a unit area of the wall per unit time.
- An increase of temperature causes an increase in pressure (the Law of Gay-Lussac). KE of the molecules increases with T. Faster molecules cause an increase in pressure for two reasons:
  - There are more collisions with the wall per unit time;
  - Each collision exerts a greater force, since the molecule is moving faster.
- An increase in the amount (moles) of gas causes an increase in pressure at constant T (Avogadro's hypothesis). More gas means larger N/V. Higher N/V means more collisions per unit time and higher pressure.
- The ratio of diffusion rates of two gases is proportional to the square root of the inverse ratio of their molar masses (Graham's Law). This is a consequence of the average kinetic energy of a gas molecule being dependent only on T. At a particular T, a heavy molecule moves slower than a light one, but its larger mass offsets its smaller speed, giving the same kinetic energy. An important consequence of this is that a sample of gas with heavy molecules will exert the same pressure as a sample of gas with the same number of light molecules, as long as their temperatures and volumes are the same. Both samples satisfy the ideal gas law, independent of what their molecules are like. Although heavy molecules move slowly and strike the wall less often, they strike it with bigger mass, and cause the same pressure.

The Historical Role of Gas Phase Chemical Reactions

The first decade of the 19th century must have been both exhilarating and frustrating for chemists. On the one hand, Dalton's atomic theory (1803) rationalized many known facts about chemical elements, compounds, and reactions. In particular, the theory provided an explanation for the laws of conservation of mass, definite proportions, and multiple proportions, and gave a rational basis for the idea of reproducible combining masses of the elements in compound formation (*e.g.*, the combining masses of oxygen and hydrogen in forming water are in the ratio 7.94 to 1).

On the other hand, attempts to develop a scale of atomic masses were thwarted by the ignorance of chemical formulas. Chemists, led by Dalton, assumed that the formula for water was HO, and that water was formed from hydrogen and oxygen by equation:

$$\mathrm{H + O \rightarrow HO}$$

This was the simplest assumption to make in the absence of definite knowledge of the formulas of elemental hydrogen, elemental oxygen, and water.

In 1809, Joseph Gay-Lussac conducted a series of experiments with gases that ultimately enabled the development of the atomic mass scale. Specifically, for reactions in which gases react to form other gases, he studied the relationships among the volumes of gaseous reactants consumed and the volumes of gaseous products formed. His result is of tremendous importance.

Gay-Lussac discovered that, as long as the experiments were conducted at a constant temperature and pressure, the volumes of reactants used and products produced always gave whole-number ratios. One of many reactions that he studied was that between hydrogen and oxygen to form water vapour. He found consistently and reproducibly that for every one volume of oxygen used, two volumes of hydrogen were required, and two volumes of water vapour were produced:

$$\text{2 vol hydrogen + 1 vol oxygen} \rightarrow \text{2 vol water vapour}$$

Gay-Lussac was not certain what to make of these results; however, Amadeo Avogadro was. He made the (to him) reasonable proposal that Gay-Lussac's results implied that equal volumes of gases, at the same temperature and pressure, contain equal numbers of molecules. In modern terms, we say that the volume occupied by a gas is proportional to the number of moles of gas present.

Operating from this assumption, he concluded and stated that Dalton's view of the hydrogen/oxygen reaction, expressed in 5-4-1, was incorrect, because it was not consistent with the two-to-one-to-two hydrogen/oxygen/water volume ratios that Gay-Lussac observed. He then went on to propose that hydrogen and oxygen occur as diatomic molecules, and that water in fact contains two atoms of hydrogen per atom of oxygen, so that the water formation reaction becomes:

$$2H_2(g) + O_2(g) \rightarrow 2H_2O(g)$$

This is the simplest proposal consistent with the combining volumes observed; as we know today, it is entirely correct.

Gas Production via Double Displacement Reactions

Three broad classifications of chemical reactions: electron transfer processes, proton transfer processes, and double displacement processes. We defined a double displacement reaction as one in which the positive and negative portions of the reactant molecules interchange to form products. It is appropriate on gases to focus briefly on double displacement reactions that produce a gas as one of the products. An important series of examples of such reactions involves metal carbonate compounds (called carbonate salts) and acids as reactants. The reaction of a metal carbonate with an acid is shown generically in equation.

$$MCO_3(\text{s or aq}) + HX(aq) \rightarrow MX_2(aq) + H_2CO_3(aq)$$

This reaction does not have a gas as one of its products. However, in a subsequent step, carbonic acid, H_2CO_3, decomposes to water and carbon dioxide, a gas that escapes in part from the aqueous solution in the form of effervescence:

$$H_2CO_3(aq) \rightarrow H_2O + CO_2(g)$$

A specific example of the acid-carbonate reaction. Here the net reaction, obtained as the sum of equations, is shown.

$$Na_2CO_3 + 2\ HCl \rightarrow 2NaCl + H_2O + CO_2$$

Acid-carbonate reactionsare very important in at least two contexts. First, metal carbonates that have low solubility in water make up a substantial fraction of the earth's crust (limestone is an example). Deposits of carbonate rocks in contact with slightly acidic ground waters undergo reactions. In so doing, they serve to maintain the acidity level of the water at a fairly constant level, supporting aquatic plant and animal life.

Second, reactions are important in the context of acid rain, which results when nitric acid, HNO_3, and sulfuric acid, H_2SO_4, are produced in the atmosphere from SO_2, produced in coal combustion, and NO_2, a byproduct of the internal combustion engine. Limestone (primarily calcium carbonate) is widely used in construction, and has been used for centuries as a medium for sculptors.

Repeated and prolonged contact of building edifices and priceless works of art with acid rain results in their slow destruction as calcium carbonate reacts with nitric and sulfuric acids according to equations. Building edifices can be replaced; works of art cannot. Finally, dietary calcium tablets are made primarily of calcium carbonate. Contact of a stomach tablet with stomach acid (hydrochloric acid) causes the tablet to dissolve via reaction, after which the calcium can be utilized by the body.

$$CaCO_3(s) + HCl \rightarrow CaCl_2(aq) + H_2O + CO_2(g)$$

Metal sulfites (for example, Na_2SO_3) react with acids in analogous fashion to form aqueous sulfurous acid, H_2SO_3. This subsequently decomposes to water and sulfur dioxide:

$$Na_2SO_3(aq) + HNO_3(aq) \rightarrow NaNO_3(aq) + H_2SO_3(aq)$$

$$H_2SO_3(aq) \rightarrow H_2O + SO_2(g)$$

Example. A 500-mg calcium tablet dissolves in an excess of stomach acid. What volume of CO_2 is produced?

Solution. We begin with a balanced chemical equation for the reaction of stomach acid and the calcium tablet, which we assume to be pure calcium carbonate:

$$CaCO_3(s) + 2HCl(aq) \rightarrow CaCl_2(aq) + H_2O + CO_2(g)$$

To use the numerical information in the equation coefficients, we must work in moles:

moles $CaCO_3$ = 0.500 g/(100.086 g/mole) = 5.00×10^{-3} moles.

1 mole of CO_2 is produced for each mole of calcium carbonate that reacts. It follows that 5.00×10^{-3} moles of CO_2 are formed. The ideal gas law enables us to calculate the volume of CO_2. We assume the temperature in the stomach to be 98.6 °F, or 37 °C, and that the carbon dioxide is produced at 1 atm pressure. Then

$$V = nRT/P = (5.00\times10^{-3}\ \text{moles})(0.08206\ \text{atm-L/mole-K})(310\ \text{K})/(1.00\ \text{atm})$$

$$= 0.127\ \text{L}$$

This volume of CO_2 is probably not sufficient to cause "indigestion."

Elementary Particles of Matter

The first elementary particle, the electron, was discovered by Thompson. But elementary particle physics as a separate area of physics was really started in the 1940's and 1950's, some time after the discovery of proton, neutron, and positron, that are all elementary particles (at least they were regarded as such until about 1970). There were two important reasons for the blossoming of elementary physics, one experimental and one theoretical.

The experimental impetus was given by the observation that very high energy particles, first observed in cosmic radiation, are able to initiate reactions, similar to nuclear interactions, when collide with the nuclei of O, N, H,... of the atmosphere. It was observed, using photographic emulsions flown at high altitude, that when particles of the cosmic radiation (mostly protons) collide with nuclei they produce a large number of new particles, at very narrow forward angles, sometimes a far larger number than the number of nucleons in the target nucleus.

The large cross section for these collisons and the shear number of particles produced were a proof that the interaction between the nucleus and the incoming cosmic ray particle is very strong. Thus, it was immediately thought that most of the particles produced were not nucleons, but rather the quanta of the strong force.

The existence of these quanta, called p-mesons, was predicted on purely theoretical grounds by Hideki Yukawa in the 1930's. It was soon proved that indeed most of the particles produced of these high energy collisions are p-mesons. Their approximate mass was also predicted successfully based on the structure of Yukawa interactions,

$$V_{Yukawa} = - g^2 e^{- m\, r\, c/\, hbar/}\, r = - g^2 e^{- r/\, R}$$

where m is the mass of the particle communicating the interaction. Note that electromagnetic interactions are similar if we take $g^2 \rightarrow k\, e^2$ and m=0 (the mass of the photon is 0!). By estimating the range of nuclear interactions, which is equal to R = hbar/(m c) = hbar c/ (m c^2) one is able to come up with a number

$$mc^2 = \text{hbar } c/\ R = 197.3 \text{ MeV f}/\ 1.2 \text{ f} = 164 \text{ MeV}.$$

The true mass of p-mesons is 138 MeV, remarkably close to this crude estimate. The investigation of interactions of nucleons and p-mesons lead to the development of elementary particle physics. The other ingredient in the emergence of elementary particle physics was the success of quantum field theory that became and still is the only successful framework of describing the interactions (creation, annihilation, and scattering) of elementary particles. Quantum field theory was started by Dirac around 1930 from the marriage of relativity and quantum mechanics.

Though the theoretical framework was understood quanum field theory was useless for most practical reasons because calculations, beyond a trivial first approximation cannot be made any sense, they were plagued by divergences (infinities) that could not be eliminated. It took almost two decades until Tomonaga, Schwinger, and Feynman could device a self-consistent method of eliminating these divergences. This method is called renormalization.

The basic idea behind renormalization can be understood easiest in is the framework of Quantum electrodynamics, the quantum field theory describing the interaction of photons with charged particles, such as electrons.

Quantum field theory implies that the vacuum is not really empty, but filled with "virtual" particles that are keep being produced and annihilated. Conservation of energy momentum is satisfied because the mass-energy of these virtual particles does not equal to the mass-energies observed for physical particles traveling in the vacuum. Virtual particles live for a very short time, before annihilating, thus the uncertainty relation between time and energy requires that their mass-energy is undetermined to a large degree.

The presence of these virtual particles on real electrons is profound and twofold. The mass-energy of the real electron is changed substantially by interactions with the bath of virtual particles around it. Note that even classically the self energy of a pointlike electron is infinite. It is also infinite, due to interactions with virtual particles, in quantum field theory.

These interactions will give an infinite contribution to the mass energy of electrons. It turns out, however that if the mass-energy of the electron (the so-called bare mass-energy) without the interaction with virtual particles is chosen to be infinitely large and negative then the sum of contributions can be made finite and one can end up with a finite mass in a consistent series of approximations of arbitrary precision.

Similarly, the other parameter describing electrons, their charge, can also be treated in a similar manner. Note that the presence of the negatively charged electron polarizes virtual particles, the total charge of which must be equal to zero.

Thus, the electron sits in a polarized dielectric medium. Then, observed from practically infinite distance (that is what experimental observations practically do) the charge of the electron seems considerable smaller due to screening. In fact if, starting form infinity, one would be able to go closer and closer to the electron, one would observe a larger and larger negative charge.

One can show that if one wishes to keep the observed charge (the one observed from infinite distance) finite then the charge one would observe at zero distance from the electron (the so-called bare charge), when all the layers of screening are peeled away, would have to be infinitely large negative.

Conversely, chosing the "bare charge" to be infinite one can build up a scheme of systematic calculations that can in principle performed to arbitrary precision. Calculations performed following the method of Tomonaga, Schwinger, Feynman and Dyson had a tremendous success in explaining experiments of ultimate precision, such as the anomalous magnetic moment of electrons and the Lamb shift.

Conservation Laws

After the discovery of pions, other elementary particles were discovered. Pions, or p-mesons, are unstable. That is why we do not see them under normal circumstances. The lifetime of π^+ (or π^-) mesons is not very short, by elementary particle standards, $\tau = 2.6 \times 10^{-8}$ s.

That means, that relativistic π^+ mesons can fly to a distance of hundreds of meters in the laboratory, if we take into the effect of time dilation. In fact, these days pion beams are produced from targets at accelerators provide an everyday experimental tool.

π^+– mesons decay into m$^+$ mesons and neutrinos, $\pi^+ \rightarrow \mu^+ + \nu$. μ^+ are of course lighter than π^+, but they are not strongly interacting, only electromagnetically (they are charged!) therefore they penetrate matter readily, they can traverse thick steel slabs without interactions. This property is used for their identification. In general, when elementary particles are studies it is extremely important to identify as many of the produced particles as possible. This helps understanding the basic laws of nature.

The antiparticles of π^+ mesons are π^--mesons, they decay into $\pi^- \rightarrow \mu^- + \nu^c_\mu$ (where this latter is a muon antineutrino, as distinguished form a muon neutrino, or an electron neutrino). Muons, or m-mesons, behave much the same way as electrons (or positrons). They only have electromagnetic and weak (like beta decay) interactions.

Only their mass (106 MeV/c^2) distinguishes them from electrons. It is still not understood why nature needs to duplicate electrons in this fashion. Along with ν_μ they carry an additive quantum number, muon number that has been observed to conserve in all interactions. The electron number carried by electrons and electron neutrinos is also conserved (positrons and antineutrinos carry electron number +1). The muons decay as

$$\mu^+ \rightarrow e^+ + \nu_\mu + \nu_e$$

in about 10^{-6}s. Muons and electrons, with their respective neutrinos are called leptons.

After the discovery of pions by Lattes, Occhialini, and Powell other particles were discovered in rapid pace. First, it was observed that near the collisions particle pairs, in the form of a narrow V diracted towards the collision centre, seemed to emerge from nothing. This was immediately interpreted as the decay of a hitherto unknown neutral particle into two positive particles. These V-events were later identified as the decay of K-mesons and hyperons. While K-mesons (m= 494 MeV/c^2) are similar to p -mesons, they do not carry a nucleon number, hyperons are different (The lightest hyperon is the L^0 hyperon of m=1116 MeV/c^2).

This is shown by their decays:

$$K^+ \rightarrow \pi^+ + \pi^0,$$

$$\Lambda^0 \rightarrow p + \pi^- \text{ or } n + \pi^0$$

As if a hyperon would have nucleon inside, as this is already indicated by its large mass. Thus, we need to expand the notion of nucleon number to baryon number (baryos=heavy in greek). Nucleons and hyperons are baryons and it is not the nucleon number, A, that is conserved but the baryon number (B). This is clearly shown in production processes. Hyperons, just like nucleons, have s=1/2, while pions and kaons are s=0 bosons. As physicists learned how to build accelerators they succeeded in creating beams of various particles (by separating the products of p-p collisions by poweful magnets). Then they observed reactions as (that is a strong interaction with large cross section as 30 mb)

$$\pi^+ + n \rightarrow \Lambda^0 + K^+.$$

Obviously in the above process baryon number is conserved but nucleon number is not. It was remarkable that reactions like were never observed.

$$\pi^+ + n \rightarrow \Lambda^0 + \pi^+$$

This required the introduction of a new quantum number, strangeness, by Gell-Mann and Pais, such that $S(\Lambda^0) = -1$, $S(K^+) = 1$. At the same time in decay processes such as K-decay or hyperon decay strangeness was not conserved. This was the first case that some quantum numbers were conserved in strong interactions and electromagnetic interactions, but not in weak interactions. These are very well distinguishable, the characteristic reaction time of one is 10^{-23} s while for the other 10^{-10} s.

Other hyperons were also discovered soon, Σ^+, Σ^- and S^0 hyperons that all had $S = -1$ (of course their antiparticles have $S = +1$), the Ξ^- and Ξ^0 hyperons that have $S = -2$. All these hyperon have $B = 1$, just like the proton and the neutron, say $B(\Xi^0) = 1$. Thus Ξ^0 is produced in a reaction like

$$K^- + p \rightarrow \Xi^- + K^+$$

Note that in this reaction both B ans S are conserved. Hyperons are all unstable, after a characteristic time of about 10^{-10} s they decay into nucleons and mesons or leptons. Example:

$$\Xi^- \rightarrow \Lambda^0 + \pi^- \text{ or } \Lambda^0 + \mu^- + \nu_\mu.$$

In decay processes (weak interactions) the strangeness can only change by only one unit. Note the second of these decay processes, it is just like beta decay. Even more similar to beta decay are two modes of L^0 decay (they happen very rarely)

$$\Lambda^0 \rightarrow p + e^- + \nu_e$$

$$\Lambda^0 \rightarrow p + \mu^- + \nu_m$$

There are other additive quantum numbers, called flavours that are similar to strangeness and are also conserved in strong interactions.

Another important quantum number conserved in strong intreractions is isotopic spin. It has an important role in nuclear physics, as well. It was noticed very early that just like nucleons (p,n) all strongly interactive particles come in multiplets having equal, or almost equal mass.

Examples are (π^-, π^0, π^+), (K^+, K^0), Λ^0 (alone), $(\Sigma^+, \Sigma^0, \Sigma^-)$, (Ξ^-, Ξ^0),... These multiplets (singulet, doublet, triplet,...) have the important property that their members have the same spin, baryon number, strangeness, and approximate mass. They have identical strong interactions.

Apart from their small mass differences they are distinguished by their charge (and consequently electromagnetic interactions). In this respect they are very similar to a system with a fixed spin that can have states with different spin projections. *E.g.* we have a s=1 system that can have s_z= –1,0, and 1. Pions, having three charge states have I = 1, so they can have I_3 = –1,0 and 1, corresponding to the states of π^{-} π^0, and π^+. Experiments showed that isotopic spin is conserved in all strong interactions. *E.g.* in the collision of a π^+ and a p the total I can either be I=1/2 or I=3/2. Then the final state products must also have either I=1/2 or I=3/2.

Nucleons have isotopic spin I=1/2, since they have 2I+1 = 2 charge states, p and n. Then isotopic spin is very relevant for nuclear physics, because nuclear physics is mostly a study of strong interactions. Then combining two nucleons can lead to I=0, or 1. The only two nucleon state is d. Then d has I=0. Nuclear forces depend strongly on isospin there are no nn or pp bound states. One could explain away the state pp by invoking the Coulomb repulsion, but there is no other good reason to forbid the nn state.

Since the introduction of isotopic spin is the recognition that p and n are just two charge states of the same particle, the Pauli exclusion principle should also be applied to them. Investigate the d from the point of view of antisymmetry of the two nucleons. The p and the n, being in I=0 state, is antisymmatric for the exchange of the p and n. (remember that s=1 is a symmetric state and s=0 is an antisymetric state). Then their wave function must be symmetric in spin, because the overall wavefunction must be antisymmetric for fermions. If the spin wavefunction is symmetric then the spin must be s=1. Indeed, the s(d)=1.

If one goes to heavier nuclei, then it is obvious that the pair ($^3_2$He, $^3_1$H) forms an isodoublet. Their collision with d can only produce an I=1/2 state in strong interactions. So

$$d + {}^3_2He \rightarrow {}^4_2He + p$$

happens because I($^4_2$He) =0. Electromagnetic interactions violate I conservation, though I_3 is still conserved.

Particle Interactions, Resonances, and Quarks

After the discovery of mesons and hyperons experimentalists found a large number of excited states of these particles as well. The typical example of these resonances is the Δ particle that has four charge states (isotopic spin, I=3/2) Δ^{++}, Δ^+, Δ^0, and Δ^- This resonance was found by Fermi in an experiment in which he bombarded protons (hydrogen) by a newly created p-meson been at the Bevatron accelerator in Berkeley. The reaction he observed was

$$\pi^- + \pi \rightarrow \Delta^0 \rightarrow \pi^- + p$$

Now the intermediate state, Δ^0 is of extremely short lifetime. In fact, t ~ 10^{-23} s. Direct observation of such a particle is impossible. The uncertainty relation between time and energy, however, implies that the energy of the Δ is uncertain to about 100 MeV. The Δ^-p cross section increases when the total energy-momentum squared $E^2 = (E_p + E_\pi)^2 - c^2(p_p + p_\pi)^2 = m_\Delta^2 c^4$ This is so because there is a new channel throgh which the scattering can happen. Due to the uncertainty of m_Δ, $\delta m_\Delta \sim$ hbar/ $(c^2 \tau)$, the maximum in the p^-p cross section is not sharp, but smeared over a region of about 100Mev in the primary energy of π^-. Thus, if we plot the cross section as a function of the energy of the pion (performing the experiment at a series of different energies) then we will see a large bump at $E^2 = m_\Delta^2 c^4$. The width of the bump provides the lifetime of the Δ particle. This particle is a baryon if you require baryon number conservation. It also turns out that the spin of this particle is s =3/2. The mass of Δ is m_Δ=1232 MeV/ c^2

As time went by many more particle resonances were discovered. These resonances can be regarded as new particles on their own right. They just happen to have large enough masses so that conservation laws allow them to decay via strong interactions into strongly interacting particles. That is the reason why their lifetime are so short. As an example the mass of Δ happens to satisfy $m_\Delta > m_\pi + m_p$ so Δ is allowed to decay into π + p. After literally hundeds of resonances with exactly the same quantum numbers as some other particles were discovered the situation became untenable.

This looked more or more as the a system of states with the excitations of the same system, like an atom or a nucleus. The existence of such excitations is a sure sign of composite nature. Clearly atoms and nuclei are composite. This gave the idea to Nambu and Gell-Mann to postulate the existence of constituents for strongly interacting particles: quarks.

The most economic choice (due to Gell-Mann) turned out to be very strange: All normal and strange strongly interacting particles in every charge mode can be constructed out of three quarks, and their antiparticles (antiquarks). The three quarks are u (up), d (down), and σ (strange) quarks. Their charge are not integer times ε.

quark	*charge*	*spin*	*Baryon number*	*Strangeness*	*Isotopic spin*	I_3
u	2/3 e	1/2	1/3	0	1/2	1/2
d	–1/3 e	1/2	1/3	0	1/2	– 1/2
s	– 1/3 e	1/2	1/3	–1	0	0

If one uses these quantum numbers then all states known at that time could be built from the combination of these quarks. Examples: $\pi^- = du^c$ (where I denote antiparticles by superscript c. Ususal notation is a bar above the u, or d, or s), p = uud, n = udd, Δ^{++} = uuu, K^- = su^c, Λ^0 = sdu, etc. Note that one gets the right spins and isotopic spins as well.

Baryons, made of three quarks, must be fermions. In the ground state (zero orbital momentum) they must have spins of s = 1/2 or s = 3/2. Indeed the nucleons have s=1/2 and the D baryons have s = 3/2. Since both of these states are made of three I=1/2 quarks (u and d) their isospin must be 1/2 or 3/2 as well. Of course, we know that I(N) = 1/2 and I(Δ) = 3/2, because they have 2 and four charge states respectively.

Anyway, the existence of Δ^{++} demands I = 3/2, because I_3 is an additive quantum number and π^{++} = uuu. Each u has I_3 = 1/2 so the total I_3=3/2. But in that case I cannot be smaller than I=3/2. Similarly, Σ^+ = uus, so, since I(s) = 0, it must have I = 1. Indeed it has three charge states. Finally, the starngeness of Ξ S(Ξ) = -2 so it must contain 2 strange quarks. Concequently, Ξ^- = ssd, Ξ^0 = ssu, they have isospin I(Ξ) =1/2, because they contain one I = 1/2 quark (u or d) and two I = 0 quarks (s). The above list conatins all ground states of baryons constructed from the three quarks u, d, and s, except sss. This state was predicted by Gell-Mann and subsequently discovered and was called Ω^-. After the identification of the ground state baryons the previously discovered resonances of higher mass and frequently higher spin were also identified as excited states of these meson and baryon states, with nonzero orbital angular momenta between quarks.

Subsequently, heavier and heavier states were investigated, mostly in $e^+ - e^-$ colliders. On theoretical grounds, the existence of another quark, the charmed quark was predicted. c- anti c bound states were found as tremendously high peaks in the $e^+ - e^-$ cross sections near total energy (in the CMS) 3 GeV. These peaks were very narrow O(100 keV), which cannot be explained any other way but by the interpretation that such a new state, so called charmonium was produced (Richter and Ting got the Nobel prize for the discovery).

The fact is that the decay is from strongly interacting particles to strongly interacting particles should be much faster (*i.e.* have larger width). QCD, the theory of strong interactions. While the u d and s quarks are light, M(c) ~ 1.5 GeV/c^2

Subsequently, and unexpectedly, two more quarks, b (bottom) and t (top) (other two Nobel prizes) were discovered in similar typ of experiments. Their masses are large M(b) = 5 GeV/c^2 and M(t) = 170 GeV/c^2 (as heavy as a A=180 nucleus). Furthermore, a third lepton family was also discovered (also Nobel prize winner), with the t (positive and negative) lepton and corresponding neutrino.

There is fairly strong (astrophysical, and accelerator physics) evidence that there are no more quarks or leptons. The known leptons and quarks can be grouped into three families. First family: d, u, e, ν_e, Second family: s, c, m, ν_μ and the third family: b, t, t, ν_τ. These are listed in order of increasing mass (except possible neutrinos, whose mass is consistent with zero). All ordinary matter, including radioactive decay products are made of the members of the first family. As we see it, the world would not be substantially different if the second and third families would not exist.

It is still a mistery why the first family is duplicated twice. The additive quantum numbers distinguishing the six quarks, charge, strangness, char, and the bottom and top quantum numbers are called flavours. They are all conserved in strong and electromagnetic interactions, but with the exception of charge they all can be violated in weak interactions.

The Standard Model

There is a substantial problem with the quark assignments of baryons. Take for example the Ω^- hyperon. Its assignment is sss and its spin is s(Ω^-) = 3/2 This spin state is completely symmetric for the exchange of the s quarks (all the three spins are parallel). Consequently the wave function of the Ω^- hyperon is a completely symmetric combination of the wavefunctions of the three s quarks. Similarly the Δ baryons have isospin I = 3/2 and spin I = 3/2, both of these states are completely symmetric, i.e the three quark wavefunctions are completely symmetric.

In fact, assuming that the three quark wavefunctions are completely symmetric one can prove that the baryon ground states formed from threee u and d quarks are exactly the observed isospin I=3/2 and spin I=3/2 state (D) and an I=1/2 and spin I=1/2 state (the nucleons). This is very troublesome because it requires the symmetrizations of wavefuncions in fermions, while according to the Pauli exclusion principle the wavefunctions should be completely antisymmetric.

Now the symmetry properties of wavefunctions and their relation to the bosonic and fermionic nature of particles is a very basic law of physics and should not be violated. This gave a lot of headache to theorists, who finally came up with the right solution. The only way to solve the situation is that the the three s quarks that make up Ω^- or the three u quarks that make up Δ^{++} differ in an additional, hitherto unknown, quantum number, so they can be antisymmetrized in that quantum number.

Gell-Mann called this quantum number colour (he also named quarks). Thus there are three colours: say red, blue, and yellow and then the wavefunctions of the three quarks inside the baryon can be completely antisymmetrized in colour, requiring that the three quarks are completely symmetrized in the remainging quantum numbers, such as spin and isospin. One can show using group theory, that such an antisymmetric combination of colour states is "colorless" does not have any colour quantum numbers. In a similar way mesons are constructed by equal combination red-antired, blue-antiblu, and yellow-antiyellow quarks, that is also a colorless state. Furthermore, one can see that the baryon and meson states of this kinds are the only colorless states one can construct from three quarks and a quark antiquark pair respectively.

Since free quarks have never been seen in nature the challange is to construct a theory of quarks that allows the existence of colorless physical states constructed from quarks only. Such a theory is quantum chromodynamics (QCD).

Quantum chromodynamics is built based on the analogy of Quantum electrodynamics, the theory of interaction of photons with charged particles. The strength of electromagnetic interactions is given by e. In field theory, so a similar "strong coupling constant," γ, is defined.

Since g is much larger than e, strong interactions are stronger than electromagnetic interactions. The role of photons is to generate interactions among charged particles. There should be particles call gluons, that create interactions among particles that have the colour quantum number. The difference is that while there is only one kind of a conserved charge in electromagnetism, in QCD there are three colours to play a similar role. Accordingly, there are much more kinds of gluons (eight) that communicate interactions among objects having colour quantum numbers. While in Quantum electrodynamics an photon can turn into an electron positron pair, gluons can turn into quark-antiquark pairs.

So there is one gluon that can turn into a blue-antired quark pair, etc. The great difference is that while photons themselves are not charged (the total charge of the electron positron pair is zero), gluons themselves carry colour. Since photons interact with charged objects only, tnot being charged, they do not directly interact with each other. At the same time gluons, carrying colour charges do have direct interactions with each other. The fact that gluons interact directly with each other has far reaching consequences in strong interactions. Remember, that the vacuum screening of charges gets stronger and stronger as one goes farther and farther away from the electron. The consequence is that the effective charge is decreasing with distance. Alternatively, using the uncertainty relation that says that large distance corresponds to low energy and short distance corresponds to high energy. The one can show that the effective charge has an energy dependence, *i.e.* at low energies a lower effective charge can be used in calculations (of cross sections, decay rates, etc.), but that increases at high energies the effective charge is larger. Now in QCD, due to the self interaction of gluons the screening situation is entirely different.

The gluons that pop out from the vacuum generate antisceening, because the like colour charges and anticharges repel each other. The end result is that at large distances (or at low momentum) the effective colour charge (g) increases, while at short distances (high momentum) the effective colour charge, g decreases. One can also show that in the limit of of infinite momentum (or energy) the effective charge vanishes. This property is called asymptotic freedom.

If the effective charge is very small at high energies then the basic calculational method of field theory of expansion in a power series of charge works very well. Indeed very highenergy experiments confirm this they agree completely with the predictions of QCD, though the precision is far less than the experimental agreement with predictions of QED.

At low energies the effective charge is large and the calculations break down. The only results at low energies are from crude numerical symulations that confirm the property of confinement: all states that carry a colour quantum number (single quark, single gluon, diquark) are non existent in the free form, quarks and gluons are permanently confined inside mesons and baryons.

The picture of strongly interactive particles is then the following: If we bombard a proton with very high energy probes then it behaves as a collection of three free quarks. This can be clearly seen in experiements. The very high energy probes (think of the uncertainty relation again) probe the proton at very short distances. So if the quarks are close they interact very little. Then when we try to separate them their interaction is getting stronger. When they are at a larger distance, the attractive energy between themm increases proportionally to the distance. To separate them to infinite distance we must supply infinite energy. Thus, they are permanently confined inside mesons and baryons.

If quarks are never seen in nature in the free form the question can be raised whether they exist at all or they are some kind of a mathematical concept only, a mnemonic that tells you how to construct elementary particles. There is extensive evidence for the existence of quarks. One is the bombardment of protons by very high energy electrons. Electrons do not participate in strong interactions, only in electromagnetic and weak interactions.

Negelecting weak interactions, the scattering is purely electromagnetic. At very high energy the proton looks like a collection of free quarks. Scattering on free quarks can be easily calculated (something like Rutherford scattering) and the calculated cross sections agree precisely with the experimentally observed ones. More direct and specttacular evidence comes from high energy $e^+ - e^-$ colliders. When these particles collide in th eCMS (which is the laboratory) they create a virtual photon at rest in the laboratory, but having an enormous mass. The photon (if it is not polarized) loses the directionality of the original e^+ and e^- comlpetely. Then it is spectacular to see that the large number of strongly interacting particles produced and observed in the final state come out in two narrowly focused jets in oppositie direction.

Without quarks one would expect a more or less isotropic distribution for these produced particles. Now how do quarks explain the existence of jets? At very high energy quarks interact with the photon (they are charged!) but their strong interactions are "weak" so the only thing that happens is that the virtual photon can decay into quark-antiquark pairs, which, due to energy-momentum conservation fly out in opposite directions with very high momentum. As long as the quarks are not very far from each other they continue to fly and have still weak interactions.

When the distance between them is increasing the potential energy also increases and try to stop them. Instead of stopping them virtual quark -antiquark pairs and gluons are produced from the vacuum and join the initial quarks in a way that they become colorless object that are now free to go out to infinite distance. Since this happens after the original quarks were produced their directionality is conserved and two jets are produced. More precise calculations in QCD confirm this scenario. There is no other known theory that could produce jets as QCD does.

Nuclear Physics

The relevant length scale for measuring nuclear size is the femtometer 1 fm = 10^{-15} meters. Physicists usually call this length one *fermi*. Nuclei vary from about one to a few fermis in radius. Recall that the Bohr radius of Hydrogen is of order 10^{-10} meters, so the nucleus is far smaller than the atom. Nuclear size was first measured by Rutherford, by noting how close α-particles came to the nucleus before the scattering ceased to be pure Coulomb repulsion (at which point they were actually hitting the nuclear surface). Despite its small size, the nucleus has about 99.97% of the mass of the atom (a bit less for hydrogen).

What is the Nucleus Made of?

The simplest nucleus, that of hydrogen, is a single proton: an elementary particle of mass about 940 MeV, carrying positive charge exactly opposite to the electron's charge, having a spin of one half and being a fermion (so no two protons can be in the same quantum state).

The next simplest nucleus, called the *deuteron*, is a bound state of a proton and a neutron. The neutron, like the proton, is a spin one-half fermion, but it has no electric charge, and is slightly heavier (by 1.3 MeV) than the proton. The binding energy of the deuteron (analogous to the 13.6 eV for the Hydrogen atom) is 2.2MeV. A photon of this energy could "ionize" the deuteron into a separated proton and neutron. However, it is not necessary to actually do this experiment to establish how tightly the deuteron is bound. One need only weigh the deuteron accurately. It has a mass of 1875.61MeV. The proton has a mass of 938.27MeV, the neutron 939.56MeV, so together (but some distance apart!) they have a mass of 1877.93MeV, 2.2MeV *more* than the deuteron. Thus, when a proton and a neutron come together to form a deuteron, they must unload 2.2MeV of energy, which they do by emitting a photon (called a γ-ray at these energies).

Both protons and neutrons, being fermions, obey the exclusion principle, two protons with spin up cannot be in the same state, although two with opposite spin directions could, and a proton and a neutron can occupy the same spot at the same time! Protons and neutrons are referred to as *nucleons*. The total number of nucleons in a nucleus is usually denoted by A, where $A = Z + N$, Z protons and N neutrons. The chemical properties of an atom are determined by the number of electrons, the same as the number of protons Z. This is called the *atomic number*. Nuclei can have the same atomic number, but different numbers of neutrons. These nuclei are called *isotopes*, the Greek for "same place", since they are in the same place in the periodic table.

These nucleons attract each other with a short range but very strong force, called the *nuclear force*. The situation here is different from that for electrons in the atom, where the strong central force tends to dominate. In the nucleus the nucleons are attracted mainly by their immediate neighbours. Nevertheless, it is a useful beginning to think of this attractive force as being a potential well, as seen by an individual nucleon, and think in terms of the nuclei as filling the lowest available quantum states in this well, just as we did for electrons in the atom. We find, for example, that the Helium nucleus, $2p+2n$, (a.k.a. the α-particle) is tightly bound—the four nucleons can all occupy the lowest state in the well. However, some larger nuclei, like C, O, Fe are actually a little more tightly bound even than He (about 8.5 MeV per nucleon as opposed to about 7.5 for He) because each nucleon is attracted to its close partners, and there are more close partners in these larger nuclei. It has also been argued that some of these higher nuclei strongly resemble bound states of α-particles.

The total binding energy (usually expressed per nucleon) of any nucleus is easy to find—just as for the deuteron above, the mass of the nucleus is found accurately, and subtracted from the sum of the masses of the separate nucleons.

Coulomb Repulsion and α-decay

As one goes to really large nuclei, the binding per nucleon actually decreases. This is a consequence of the electrostatic repulsion between the protons. The essential point is that the Coulomb repulsion is *long range*, but the nuclear attraction is short range. Considering a proton at the surface of a nucleus, it is bound to the nucleus by the attractive forces of its nearest neighbours, say five or six of them, and this number is little affected as one goes up the periodic table from, say, Fe to U. On the other hand, the proton on the surface is being pushed away by the electrostatic repulsion from the other protons, and this increases substantially as the number of *other* protons goes from 25 in iron to 91 in uranium, even though they may be a little further away on average. In fact, this is why there is a limit on the size of atoms.

A large nucleus does not, however, tend to spit out protons—there's a much more economical way to unload charge. It ejects an α-particle. The binding energy of the nucleons in an α is almost exactly the same as it is in a U nucleus, so little nuclear binding has to be traded for lowering the electrostatic potential energy.

Thus the most common form of Uranium, U^{238} (A=238, Z=92) goes to Thorium (A=234, Z=90) by α-decay. The surprising thing is, it takes a time comparable to the age of the universe to do so! If it's energetically desirable, why does it take so long?

The answer is—there's a barrier in the way. It is a good model to imagine the U to have the a particle inside it, confined to a deep square well potential representing the attractive nuclear force. However, the α-particle wavefunction will penetrate some distance into the wall of this finite square well potential, where the α will be away from the nuclear force, but not away from the long-range electrostatic repulsion of the other protons in the nucleus. The picture is, then, that outside the well, the ground is not flat it's like a well at the top of a hill.

The implication of this is that if the energy level of the a inside the well is higher than the flat land surrounding the hill, the α-wavefunction after dying away underground as it tunnels out from the well will re-emerge at some distance, oscillating. In other words, there is some finite probability of finding the α shooting away from the nucleus some distance away.

The actual probability of this happening depends on how much the wavefunction decayed as it tunneled underground away from the well, which depended on how far its energy was below the ground. For U^{238}, this decay on tunneling away from the well is so rapid that the α takes on average billions of years to escape. However, many different nuclei undergo α decay, the quickest in millionths of a second. The energies of the emitted α's can be measured, and confirm this explanation of quantum mechanical tunneling, first put forward in 1928. The quickest decays correspond to the most energetic α's, just as the tunneling picture would suggest.

β-decay

These heavier nuclei have more neutrons than protons, because the Coulomb repulsion makes it harder to bind protons. However, a price is paid—thinking of a shell-type model, the extra neutrons go into higher energy shells. Thus for nuclei of a given size, there is an optimum ratio of neutrons to protons, close to one for small nuclei where the Coulomb effects are small enough to be neglected, close to 1.5 for the largest nuclei.

Note that if a large nucleus decays by emitting an α, the nucleus left behind has an even greater ratio of neutrons to protons, since $\alpha = 2p+2n$. This implies that it might be able to go to a lower energy state if it could trade a neutron for a proton. In fact, this can sometimes be done. The neutron itself is not a stable particle. A neutron alone in space will only last about ten minutes before ejecting an electron, leaving a proton behind. This is energetically possible, because the mass of the neutron exceeds the sum of the masses of the proton and electron.

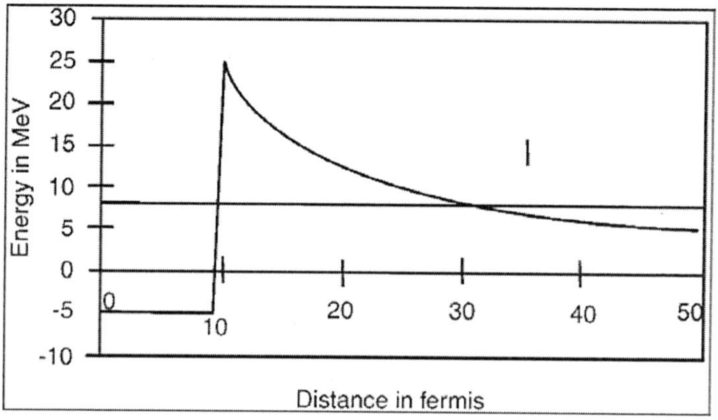

Figure: Distance in Fermis

The process is called β-decay. It can also occur in a nucleus, but now the different binding energies of the initial neutron and the final proton must be factored in to decide if it will go. Other things being equal, the more neutrons there are present in the nucleus, the more likely it is to β-decay.

Actually,α-decay was a big puzzle when it was first carefully observed, because it was discovered that identical nuclei β-decaying to the same final states ejected electrons over a range of energies, up to a maximum corresponding to the energy difference between the initial and final nuclei. This led to the suggestion that maybe, after all, energy wasn't conserved in every process in nature.

But physicists are a conservative lot when it comes to energy conservation, and Pauli suggested that maybe $n \rightarrow p+e$ wasn't the whole story some other particle was being created at the same time, carrying away the balance of the energy, but *undetected.* This meant it had no charge, no mass, and, unlike the photon, no electric interaction. Fermi called it the *neutrino*, the little neutral one.

Another reason the neutrino seemed necessary was that the initial neutron had spin one-half, as did the final proton *and* the emitted electron. It was difficult to see how angular momentum could be conserved if no other particle were emitted, because even if the final proton and electron had some orbital angular momentum, that comes in whole-number amounts, so how could two half-integer spins and an integer angular momentum add to be equal to the initial half-integer spin? It turned out they couldn't, the neutrino really was there all along, it had spin one-half itself, but it was years before it was detected.

Detecting neutrinos is still a difficult business, since they are not sensitive to the nuclear (strong) force, nor to electromagnetic fields. The force that causes neutron (and other particle) decay is called the weak force.

The only way to detect a neutrino is if it collides with a particle and interacts weakly essentially reversing one of these decays. The problem is that a neutrino moving downwards is extremely unlikely to interact with a nucleus, it will almost certainly just pass through the earth.

Therefore, to detect neutrinos takes large detectors and a strong flux of neutrinos. Much effort has been expended detecting neutrinos from the sun, since they are our only window into the nuclear processes taking place deep in the sun—they can get out unscathed, unlike photons or any other particle emitted. In fact, it is found that the number detected is less by a factor of two or three from that predicted using standard analyses of the solar nuclear processes.

It turns out that in fact the neutrino is not quite massless, and there are actually three kinds of neutrinos! Furthermore, as a neutrino passes through matter, there is a nonzero probability of it changing from one kind to another. The solar nuclear processes generate what are called electron neutrinos, and the initial efforts at neutrino detection looked for that kind.

More recent detectors count the other two kinds (mu neutrinos and tau neutrinos) as well, and the total number of neutrinos detected agrees well with the prediction of the total number generated in the sun, so it is now believed that some of the original electron neutrinos change types traveling from their point of origin to the sun's surface. Anyway, back to the radioactive decay of U^{238}. It emits an α to become Th^{234}, then β decays twice in succession to become U^{234}. This is now relatively proton-rich, and in fact a decays five times in succession to become a form of lead,

$$U^{234} \rightarrow Th^{230} \rightarrow Ra^{226} \rightarrow Rn^{222} \rightarrow Po^{218} \rightarrow Pb^{214}$$

Pb^{214} undergoes two β decays, an α decay and two more β decays before it reaches stability at Pb^{206}. The sequence above is through metallic elements with one exception—Rn, radon, is a gas because, atomically speaking, it has completely filled electron shells, and so is chemically unreactive, and gaseous at ordinary temperatures. It emits its α particle in a period of days, which gives it time to diffuse out of the rock into somebody's basement.

γ-decay

The three forms of radiation emitted by decaying nuclei, α, β and γ, were named before the particles being emitted were fully identified. It was only realized later that the β-rays were electrons. The γ rays are just high-energy photons, of order 100 keV to a few MeV.

Emission of γ rays is similar (but at much higher energies) to emission of photons by excited states of atoms. The nucleus can be excited by having just emitted an α or β, or by colliding with another nucleus, or being bombarded by neutrons, say. All these events can lead to a nucleus in which the charge distribution is oscillating, and electromagnetic radiation ensues.

The Droplet Model

In some ways, the binding of nucleons in a nucleus is reminiscent of that of molecules in a drop of water—the force is attractive between near neighbours. Notice how different this is from the electrons in an atom, held in place by a large central attracting object. This "droplet" model of the nucleus is useful for understanding nuclear fission. If the droplet is somehow caused to vibrate vigorously, it might break into two, or more, smaller nuclei.

For the U^{235} nucleus (only 0.7% in naturally occurring uranium), if a slow neutron gets too close, the attractive nuclear force pulls it towards the nucleus so strongly that in the resulting collision, the nucleus breaks into two smaller ones, and two or three neutrons are ejected (since the smaller nuclei formed have lower proportions of neutrons). There is also substantial energy release—about 200 MeV. The much more common U^{238} nucleus also attracts free neutrons, but not so strongly, and it survives neutron capture without fission.

A neutron has to be moving very slowly to be captured by U^{235}. On the other hand, when U^{235} fissions the emitted neutrons are moving very fast. To get a *chain reaction* in a nuclear reactor, the main problem is to slow down the emitted neutrons before they leave the reactor.

If they can be slowed down successfully, neutrons emitted by one fission will initiate further fissions, giving continuous—perhaps increasing—energy output. However, the neutrons bounce off the heavy uranium nuclei with almost no energy loss—to get a particle to lose energy in a collision, it must collide with something approximately its own size.

Water is often used, the neutrons lose energy bouncing off the protons in the hydrogen atoms. Unfortunately, the neutrons often stick to the protons, forming deuterons, so to keep the reaction going the uranium must be enriched—the U^{235} content boosted to one to four per cent. Of course, care must be taken to ensure that the chain reaction isn't *too* successful moderators, which absorb neutrons, are put into the reactor if the rate of fission increases beyond the planned level. Needless to say, moderators are not used in military applications of this concept.

Chapter 3

Fundamental Forces

Particles and Waves

From Bohr's Atom to Electron Waves

Reactions to Bohr's Model: Bohr's interpretation of the Balmer formula in terms of quantized angular momentum was certainly impressive, but his atomic model didn't make much mechanical sense, as he himself conceded. For example, an electron jumping from the nth orbit to the mth emitted radiation at frequency equal to the energy difference of the orbits divided by h. Presumably, it began radiating as soon as it left its original orbit.

As Rutherford put it in a letter to Bohr, "how does an electron decide what frequency it is going to vibrate at when it passes from one stationary state to another? It seems to me that you would have to assume that the electron knows beforehand where it is going to stop."

A few days later, Bohr sent Rutherford another manuscript containing the correspondence principle argument gone. This was a bit too long for Rutherford. He responded: "As you know, it is the custom in England to put things very shortly and tersely in contrast to the Germanic method, where it appears to be a virtue to be as long-winded as possible". The Germans themselves were very skeptical of Bohr's model. Harald Bohr wrote to his brother from Göttingen in the fall of 1913 that the young physicists there considered Bohr's model too "bold" and "fantastic".

Mysterious Spectral Lines

One good reason for the Germanic skepticism was the recent discovery of apparently new spectral lines for hydrogen corresponding to *half*-integers in the Balmer formula. These new lines had been seen by a spectroscopist, Alfred Fowler, in a discharge tube containing a mixture of hydrogen and helium, and also in the spectra of a star by Pickering.

When he heard about these new lines, Bohr realised that his formula for the Rydberg constant,

$$R_H = \left(\frac{1}{4\pi\varepsilon_0}\right)^2 \cdot \frac{2\pi^2 me^4}{ch^3}$$

should also apply to the ionized helium atom-that is, a helium nucleus with a single electron in orbit around it, provided e^2 is replaced by $2e^2$, to correct for the doubly charged nucleus. This gives an overall factor of 4, exactly equivalent to replacing the integers in the denominator by half-integers. In other words, Bohr argued, the supposed new hydrogen lines in fact were from ionized helium.

Fowler, however, was a very precise experimentalist. He could measure the spectral lines to five significant figures. Therefore, despite uncertainties in the values of e, m and h, the ratio of the "Rydberg constant" for helium to that for hydrogen could be measured to great accuracy. Fowler found it wasn't 4, but actually 4.0016.

Bohr's response to this news was conclusive. He pointed out that his analysis of the circular motion had neglected the finite mass of the nucleus. He should really have taken the electron to have an effective mass equal to $mM/(m + M)$. It is easy to verify that if this correction is made for both hydrogen and helium, the ratio changes from 4 to 4.0016!

This discovery won over most physicists to the point of view that Bohr was definitely on to something. Sir James Jeans remarked that the only justification for Bohr's postulates was "the very weighty one of success". According to Rosenfeld (Introduction to *Niels Bohr Collected Works*):

"At Göttingen, that high place of mathematics and physics, where the sense for propriety was strong, the prevailing impression was one of scandal, or at least bewilderment, before the undeserved success of such high-handed disregard of the canons of formal logic." But all Germans didn't feel that way. Bohr's friend George Hevesy told Einstein about Bohr's identification of the Pickering-Fowler lines with helium. Einstein said: "This is an *enormous achievement*. The theory of Bohr must then be right."

A Periodic Table Puzzle and the X-ray Connection

Mendeleev's success in demonstrating that the elements had recurring patterns of chemical behaviour when arranged in order of increasing atomic weight had some anomalies. In particular, cobalt has a higher atomic weight than nickel, yet its chemical properties strongly suggested it should come *before* nickel in the table. It was gradually realised over the period from 1907 to 1913 (primarily by a Dutchman, van den Broek, actually a real estate lawyer who did physics in his spare time) that the significant parameter from a chemical point of view was not the mass number A, but the total number of electrons, and hence the charge on the nucleus Z, once the nuclear model was accepted.

The problem was measuring the nuclear charge. This could in fact be done by Rutherford scattering, but that was a long and tedious process. X-rays gave a quicker way. Any substance on being bombarded with sufficiently fast electrons emits a continuum of x-ray frequencies up to a maximum frequency f given by hf = kinetic energy of electron, *plus some sharply defined lines*. The frequency of the sharply defined lines gradually increases with atomic number.

These x-ray lines were first discovered by Max von Laue in Munich in the summer of 1912. Harry Moseley, who worked with Rutherford mostly teaching undergraduates, decided to investigate this new spectroscopy. When Bohr came to visit in 1913, he encouraged Moseley to investigate x-ray spectra of many elements with a view to pinning down the values of Z. Moseley used a cathode ray tube a yard long with a little train inside it having a different element in each car, so many readings could be taken in one run. He found several different x-ray lines for each element, with very similar patterns for each element, although gradually shifting to higher frequencies with increasing Z. For example, one particular line, labeled $K\alpha$, had a Z-dependent frequency:

$$f(Z) = R_H(Z-1)^2\left(\frac{1}{1^2}-\frac{1}{2^2}\right)$$

This made it possible to order all the elements unambiguously. It was also good evidence that Bohr's theory was close to correct for the innermost electrons in the atom, those in the field of the unshielded nucleus. Actually, the appearance of Z–1 rather than Z in the formula is because the innermost shell has just two electrons, and they partially shield the nuclear charge from each other. The appearance of this factor was not understood at the time, because it was thought that there were more electrons in the innermost shell. Bohr later remarked that Moseley's extremely clear x-ray work had far more impact than Rutherford's scattering experiment in widening acceptance of the Bohr-Rutherford atom. Tragically, Moseley died two years later, killed by a bullet in the head at Gallipoli.

What Bohr's Model Achieved

As Bohr himself emphasized, his model atom didn't *explain* anything, in the way classical dynamics or electromagnetism could explain how systems functioned in terms of a few underlying laws. What the Bohr model did do was to tie together previously unrelated phenomena, and therefore clarify the agenda for future investigations. It showed there was a link between the Balmer formula, Planck's constant and the nuclear atom, with quantized angular momentum playing a central role.

The precision of the spectral predictions for *ionized* helium demonstrated something was right about the model, as did the interpretation of characteristic x-rays for many elements in terms of inner-shell electrons.

On the other hand, despite heroic efforts over several years by Bohr and others, there was no successful prediction of the observed spectral lines for the next simplest atom neutral helium. Evidently, to make further progress in understanding atoms, some new insight was needed.

At this point the war intervened. Much of the momentum in physics was lost. Rutherford's lab was practically emptied, Rutherford himself spent most of his time on antisubmarine research. Bohr put much effort into establishing a physics institute in Copenhagen. He also thought at length about how the periodic table might be understood in terms of filling an atom with electrons.

There was something of a breakthrough in 1924 when the German physicist Wolfgang Pauli suggested there must be a law that two electrons couldn't go into the same orbit. (By this point, the number of allowed orbits had been substantially increased by Sommerfeld to include Keplerian elliptic ones.) Pauli's "exclusion principle" was introduced as an empirical rule. We shall see later that it follows very naturally from symmetry considerations. We shall also find that in the full quantum picture of the atom, although the angular momentum is quantized, as Bohr realised, the orbital angular momentum is actually *zero* in the ground state of the hydrogen atom.

One of the more surprising results of the full quantum theory is that the energy levels found by Bohr, and even relativistic corrections computed by Sommerfeld, are exactly correct, even though the model of classical inverse square orbits is far from the truth (except for very large orbits).

A Student Prince Catches a Wave

Despite all the effort in Copenhagen and Göttingen, the next real advance in understanding the atom came from an unlikely quarter a student prince in Paris. Prince Louis de Broglie was a member of an illustrious family, prominent in politics and the military since the 1600's.

Louis began his university studies with history, but his elder brother Maurice studied x-rays in his own laboratory, and Louis became interested in physics. He worked with the very new radio telegraphy during the war.

After the war, de Broglie focused his attention on Einstein's two major achievements, the theory of special relativity and the quantization of light waves. He wondered if there could be some connection between them. Perhaps the quantum of radiation really should be thought of as a particle. If so, the theory of special relativity suggested that the observed energy-momentum relationship, $E = cp$, meant that it had a very small rest mass, for it was always observednaturallyto be moving at the speed of light.

De Broglie suspected that if the speed of a sufficiently low energy quantum could be measured, it would be found to be less than c. On this point he was wrong (as far as we know!). Nevertheless, it was a very valuable conceptual breakthrough to think of the quantum of radiation as a *particle*, knowing full well that radiation is a wave.

In fact, his incorrect idea that the photon (as we now call the light quantum) had a rest mass led him to analyse the relationship between particle properties and wave properties by transforming to the rest frame of the photon, and he discovered that the energy and momentum of the particle were related to the frequency and wavelength of the wave by:

$$E = hf,\ p = h/\lambda$$

Of course, the first condition is the Planck-Einstein quantization, and the second follows trivially from it if we take $E = cp$ and $\lambda f = c$. But de Broglie showed it was more generally true it worked even if the photon had a rest mass. Having decided that the photon might well be a particle with a rest mass, even if very small, it dawned on de Broglie that in other respects it might not be too different from other particles, especially the very light electron.

In particular, maybe the electron also had an associated wave. The obvious objection was that if the electron was wavelike, why had no diffraction or interference effects been observed?

But there was an answer. If de Broglie's relation between momentum and wavelength, $p = h/\lambda$, also held for electrons, the wavelength was sufficiently short that these effects would be easy to miss. As de Broglie himself pointed out, the wave nature of light isn't very evident in everyday life, or in ray tracing in geometrical optics.

He suspected the apparently pure particle nature of electronic trajectories was analogous to the apparent straight line propagation of rays of light, over distance scales much greater than the wavelength.

However, the wavelike properties should be important on an atomic scale. No progress had been made in a decade in understanding why the electronic orbits in the Bohr atom were restricted to integral values of the angular momentum in units of h. But if the electron were in some sense a wave, it would be very natural to restrict the orbits to those of standing waves, for otherwise the electron wave on going around the orbit would interfere with itself destructively.

Suppose now the electron, having momentum p, is moving in a circular orbit of radius r. Then for a standing wave, a whole number of wavelengths must fit around the circle, so for some integer n, $n\lambda = 2\pi r$. Putting this together with $p = h/\lambda$, we find:

$$2\pi r = n\lambda = nh/p$$

so,

$$L = pr = nh/2\pi.$$

The "standing wave" condition immediately gives Bohr's quantization of angular momentum! This was the prince's Ph. D. thesis, presented in 1924. His thesis advisor was somewhat taken aback, and wasn't sure if this was sound work. He asked de Broglie for an extra copy of the thesis, which he sent to Einstein. Einstein wrote shortly afterwards: *"I believe it is a first feeble ray of light on this worst of our physics enigmas*".

An Accident at the Phone Company Makes Everything Crystal Clear There was an accident at the Bell Telephone Laboratories in April 1925. Clinton Davisson and L. H. Germer, looking for ways to improve vacuum tubes, were watching how electrons from an electron gun in a vacuum tube scattered off a flat nickel surface. Suddenly, while the experiment was running and the nickel target was very hot, a bottle of liquid air near the apparatus exploded, smashing one of the vacuum pipes, and air rushed into the apparatus.

The hot nickel target oxidized immediately. The layer of oxide made their target useless for further investigations. They decided to clean off the oxide by heating the nickel in a hydrogen atmosphere then in vacuum. After doing this for a prolonged period, the nickel looked good, and they resumed the investigation. To their amazement, the pattern of electron scattering from the newly cleaned nickel target was completely different from that before the accident. What had changed? On examining their newly cleaned crystal carefully, they found a clue. The original target was polycrystalline made up of a multitude of tiny crystals, oriented randomly. During the prolonged heating of the cleaning process, the nickel had re-crystallized into a few large crystals.

To quote from their paper: "It seemed probable to us from these results that the intensity of scattering from a single crystal would exhibit a marked dependence on crystal direction, and we set about at once preparing experiments for an investigation of this dependence. We must admit that the results obtained in these experiments have proved to be quite at variance with our expectations.

It seemed likely that strong beams would be found issuing from the crystal along what may be termed its transparent directions the directions in which the atoms in the lattice are arranged along the smallest number of lines per unit area. Strong beams are indeed found issuing from the crystal, but only when the speed of bombardment lies near one or another of a series of critical values, ant then in directions quite unrelated to crystal transparency.

"The most striking characteristic of these beams is a one to one correspondence...which the strongest of them bear to the Laue beams that would be found issuing from the same crystal if the incident beam were a beam of x-rays. Certain others appear to be analogues... of optical diffraction beams from plane reflection gratings the lines of these gratings being lines or rows of atoms in the surface of the crystal.

Because of these similarities... a description... in terms of an equivalent wave radiation... is not only possible, but most simple and natural. This involves the association of a wavelength with the incident electron beam, and this wavelength turns out to be in acceptable agreement with the value h/mv of the undulatory mechanics, Planck's action constant divided by the momentum of the electron. "That evidence for the wave nature of particle mechanics would be found in the reaction between a beam of electrons and a single crystal was predicted by Elsasser two years agoshortly after the appearance of L. de Broglie's original papers on wave mechanics."

Wave Packets

The Wave-Particle Puzzle

De Broglie's explanation of the Bohr atom quantization rules, together with the accidental discovery of electron diffraction scattering by Davisson and Germer, make a very convincing case for the wave nature of the electron. Yet the electron certainly behaves like a particle sometimes.

An electron has a definite mass and charge, it can move slowly, it can travel through a piece of apparatus from a gun to a screen.

What, then, is the relationship between the wave and particle viewpoints? De Broglie himself always felt both were always present. He called the wave a pilot wave, and thought it guided the motion of the particle. Unfortunately, that viewpoint leads to contradictions.

The standard modern interpretation is that the intensity of the wave (measured by the square of its amplitude) at any point gives the relative probability of finding the particle at that point. This interpretation, originally presented by Max Born in 1926, is parallel to the relation between the electromagnetic field and quanta the probability of finding a quantum (photon) at any point is proportional to the energy density of the field at that point, which is the square of the electric field vector plus the square of the magnetic field vector.

The standard notation for the de Broglie wave function associated with the electron is y (x,t). Thus, $|\text{y}\,(x,t)|^2 Dx$ is the relative probability of finding the electron in a small interval of length Dx near point x

at time t. (For the moment, we restrict the electron to move in one dimension for simplicity. The generalization is straightforward.)

Keeping the Wave and the Particle Together?

Suppose following de Broglie we write down the relation between the "particle properties" of the electron and its "wave properties":

$$\tfrac{1}{2}mv^2 = E = hf, \quad mv = p = h/\lambda$$

It would seem that we can immediately figure out the speed of the wave, just using 1 $f = c¢$, say.

We find:

$$l\,f = (h/mv).\,(\tfrac{1}{2}mv^2/h) = \tfrac{1}{2}v.$$

So the speed of the wave seems to be only half the speed of the electron! How could they stay together? What's wrong with this calculation?

Localizing an Electron

To answer this question, it is necessary to think a little more carefully about the wave function corresponding to an electron traveling through a vacuum tube, say. The electron leaves the cathode, shoots through the vacuum, and impinges on an anode of a grid. At an intermediate point in this process, it is moving through the vacuum and the wave function must be nonzero over some volume, but zero in the places the electron has not possibly reached yet, and zero in the places it has definitely left. However, if the electron has a precise energy, say fifty electron volts, it also has a precise momentum. This necessarily implies that the wave has a precise wavelength. But the only wave with a *precise* wavelength 1 has the form,

$$\psi(x,\ t) = A\ \sin\ (kx–\omega t)$$

where $k = 2\pi/\lambda$, and $\omega = 2\pi f$. The problem is that this plane sine wave extends to infinity in both spatial directions, so cannot represent a particle whose wave function is non zero in a limited region of space.

Therefore, to represent a localized particle, we must superpose waves having different wavelengths. The principle is best illustrated by superposing two waves with slightly different wavelengths, and using the trigonometric addition formula:

$$\sin((k - \Delta k)x - (\omega - \Delta\omega)t) + \sin((k + \Delta k)x - (\omega + \Delta\omega)t)$$

$$= 2\sin(k - \omega k)\cos((\Delta k)x - (\Delta\omega)t)$$

This formula represents the phenomenon of beats between waves close in frequency. The first term, sin(kx-w t), oscillates at the average of the two frequencies. It is modulated by the slowly varying second term, which oscillates once over a spatial extent of order $\pi/\Delta k$. This is the distance over which waves initially in phase at the origin become completely out of phase. Of course, going a further distance of order $\pi/\Delta k$, the waves will become synchronized again.

That is, beating two close frequencies together breaks up the continuous wave into a series of packets, the beats. To describe a single electron moving through space, we need a single packet.

This can be achieved by superposing waves having a continuous distribution of wavelengths, or wave numbers within of order Δk, say of k. In this case, the waves will be out of phase after a distance of order $\pi/\Delta k$, but since they have many different wavelengths, they will never get back in phase again. (This is all covered properly in the theory of Fourier transforms. Here we are just trying to make it plausible.) It should be evident from the above argument that to construct a wave packet representing an electron localized in a small region of space, the component waves must get out of phase rapidly. This means their wavelengths cannot be very close together. In fact, it is not difficult to give a semi-quantitative estimate of the spread in wavelength necessary, just from a consideration of the two beating waves.

A packet localized in a region of extent Δx can be constructed of waves having k's spread over a range Δk, where $\Delta x \sim \pi/\Delta k$.

Now, $k = 2\pi/l$, and $p = h/\lambda$, so $k = 2\pi\ p/h$.

Therefore, $\Delta k = 2\pi\ \Delta p/h$, and $\Delta x \sim \pi/\Delta k \sim h/\Delta p$ (dropping the factor of 2).

Thus: $\Delta x \Delta p \sim h$

This is Heisenberg's Uncertainty Principle.

Phase Velocity and Group Velocity: Keeping the Wave and Particle Together

Establishing that an electron moving through space must be represented by a wave packet also resolves the paradox that the velocity of the waves seems to be different from the velocity of the electron. The point is that the electron waves, like water waves but unlike electromagnetic waves, have differing phase and group velocities.

To see this, consider again the beating of two waves of slightly different wavelengths.

$$\sin((k-\Delta k)x-(\omega-\Delta\omega)t)+\sin((k+\Delta k)x-(\omega+\Delta\omega)t)=$$
$$2\sin(kx-\omega t)\cos((\Delta k)x-(\Delta\omega)t)$$

The waves described by the term $\sin(kx-\omega t)$ have velocity $\frac{1}{2}v$, as previously derived. But the envelope, the shape of the wave packet, has velocity $\Delta\omega/\Delta k$ rather than ω/k. These velocities would be the same if ω were linear in k, as it is for ordinary electromagnetic waves. But the ω–k relationship follows from the energy-momentum relationship for the (non-relativistic) electron, $E = \frac{1}{2}mv^2 = p^2/2m$.

So $dE/dp = p/m = v$. But $E = hf = h\omega/2$p, and $p = h/\lambda = hk/2\pi$.

Therefore, $\Delta\omega/\Delta k = dE/dp = v$. So the packet travels at the speed we know the electron must travel at, even though the wave peaks within the packet travel at one-half the speed.

Probabilities, Amplitudes and Probability Amplitudes

Here we review three kinds of double slit experiments *classical particles*, *classical waves*, and *quantum things*.

Bullets

First we consider a double slit experiment with *bullets*. The source is a not very accurate machine gun, which sprays bullets towards two parallel slits, 1 and 2, in a sheet of armorplate, which may be covered one at a time with another piece of armor. The bullets are "detected" by plowing into a soft (but thick) wood screen some distance beyond the slits. We assume the bullets don't break up, you always find whole bullets in the wood.

If we now cover slit 2, and fire a lot of bullets, by examining the screen at the end we will see where the bullets went, and be able to construct a probability distribution $P_1(x)dx$ giving the probability that a bullet going through slit 1 lands in the interval x, $x + dx$. Similarly, covering slit 1, we can find the probability distribution $P_2(x)$ for bullets going through slit 2. Now suppose we open both slits this gives a probability distribution $P_{12}(x)$ and it is clear that this will be simply the sum:

$$P_{12}(x) = P_1(x) + P_2(x).$$

This of course follows from the self-evident fact that any bullet getting to the screen must have gone either through slit 1 or slit 2. *The probabilities add.* There is one other important point about bullets they are lumpy. You either find one or you don't. This is also true of photons and electrons.

Water Waves

Now think about a double slit experiment with waves, say, surface waves on water the standard ripple tank experiment. Assume that plane waves of definite wavelength impinge from the left on a barrier having two slits P_1 and P_2, with P_2 being initially closed. Then to the right of the barrier, waves emanate from the open slit with circular wavefronts. These waves are detected at a screen over to the right. The waves could be detected by having corks bobbing up and down in the water just in front of the screen.

We could at some instant in time measure the height distribution $h_1(x)$ of the corks from waves coming through slit 1 (slit 2 being closed). Of course, this height distribution is changing all the time, and is not really what interests us. What we really want to know is the intensity of the wave motion at each point on the screen (analogous to the brightness of light, for example). In any simple harmonic motion, the intensity is proportional to the square of the amplitude. For our surface waves on water, the amplitude is of course just the height of the wave (relative to still water), so the energy intensity $I_1(x)$ for waves coming through slit 1 has the form

$$I_1(x) = Ah_1^2(x).$$

What happens when both slits are open? It is well known that waves superpose linearly—that is, the *heights* add. If at some instant of time with both slits open the height distribution of the corks is given by $h_{12}(x)$, then

$$h_{12}(x) = h_1(x) + h_2(x).$$

In this harmonic motion, the corks bob downwards as much as upwards, so h_1 and h_2 are as likely to cancel as they are to reinforce each other. The energy intensity when both slits are open,

$$I_{12}(x = Ah_{12}^2(x) = I_1(x) + I_2(x) + Ah_1(x)h_2(x).$$

The crucial point here is that in contrast to the bullets, the intensity when both slits are open is *not* the sum of the separate intensities for the slits being open one at a time. This is just the familiar wave interference effect. Notice that in this classical wave picture, there is no lumpiness. The energy of the wave lapping against the screen varies smoothly, and can have any value we choose.

Classical Light

Historically, an experiment equivalent to the double slit experiment for light was what convinced people it was indeed a wave.

The light reaching a screen exhibited the identical intensity pattern observed for water waves. This was fully explained with the advent of Maxwell's equations: the light was simply an electromagnetic wave, the amplitude of the electric field being analogous to the height of the water wave in the discussion above. To find the total electric field E_{12} at any point on the screen, the electric field E_1 of the wave emanating from S_1 is added to E_2 from S_2. The intensity of light I_{12} reaching the screen is proportional to E_{12}^2. Just as for the water waves, E_1 and E_2 oscillate about mean values of zero, so the analogy is complete.

Photons

The discovery that light is after all lumpy on a fine enough scale throws this picture into confusion. Experimentally, light of frequency f can only be detected in lumps of size hf. To the classically trained mind, this naturally suggests that the source of light is actually sending out these lumps, and each lump presumably passes through one of the slits on its way to the screen. However, this naïve picture leads to a contradiction—how could it generate the observed diffraction pattern? Experimentally, the diffraction pattern builds up even if the light is so dim that there is a time interval between each photon (lump) passing through the apparatus. The picture of each photon going through one of the slits cannot be correct, because the distance between the slits governs the brightness pattern on the screen, and thus the probability pattern of where the photons land.

The only known resolution to this conceptual problem is to give up trying to visualize the path of the photon, just solve Maxwell's equation for the wave propagation, then interpret the resulting wave intensity as giving the relative probability of detecting the photon at any point. This is how quantum mechanics is done. It does not correspond to a physical picture readily interpretable in terms of familiar concepts. However, it does accord well with what is observed in nature.

Electrons

Electrons behave in the same way as photons; there is little to add to the description. The irony is, of course, that before quantum mechanics electrons were seen as bullets, light as waves, and it has turned out that they are *both* basic quantum particles, with their propagation governed by *wave* equations: but when they arrive, they act like *particles*.

The Uncertainty Principle

Waves are Fuzzy

The *wave nature* of particles implies that we cannot know *both* position *and* momentum of a particle to an arbitrary degree of accuracy if Δx represents the uncertainty in our knowledge of position, and that of momentum, then

$$\Delta p \Delta x \sim h$$

where h is Planck's constant. In the real world, particles are three-dimensional and we should say

$$\Delta p_x \Delta x \sim h$$

with corresponding equations for the other two spatial directions. The fuzziness about position is related to that of momentum *in the same direction.* Let's see how this works by trying to measure y-position and y-momentum very accurately. Suppose we have a source of electrons, say, an electron gun in a CRT (cathode ray tube, such as an old-fashioned monitor). The beam spreads out a bit, but if we interpose a sheet of metal with a slit of width w, then for particles that make it through the slit, we know y with an uncertainty $\Delta y = w$. Now, if the slit is a long way downstream from the electron gun source, we also know p_y very accurately as the electron reaches the slit, because to make it to the slit the electron's velocity would have to be aimed just right.

But does the measurement of the electron's y position in other words, having it go through the slit affect its y momentum? The answer is *yes*.

If it didn't, then sending a stream of particles through the slit they would all hit very close to the same point on a screen placed further downstream. But we know from experiment that this is *not* what happens a single slit diffraction pattern builds up, of angular width $\theta \sim \lambda/w$, where the electron's de Broglie wavelength λ is given by $px \cong h/\lambda$ (there is a negligible contribution to λ from the y-momentum). The consequent uncertainty in p_y is,

$$\Delta p_y / p_x \sim \theta \sim \lambda/w$$

Putting in $p_x = h/\lambda$, we find immediately that,

$$\Delta p_y \sim h/w$$

so the act of measuring the electron's y position has fuzzed out its y momentum by precisely the amount required by the uncertainty principle.

Trying to Beat the Uncertainty Principle

In order to understand the Uncertainty Principle better, let's try to see what goes wrong when we actually try to measure position and momentum more accurately than allowed. For example, suppose we look at an electron through a microscope. What could we expect to see? Of course, you know that if we try to look at something *really* small through a microscope it gets blurry a small sharp object gets diffraction patterns around its edges, indicating that we are looking at something of size comparable to the wavelength of the light being used.

If we look at something much smaller than the wavelength of light like the electron we would expect a diffraction pattern of concentric rings with a circular blob in the middle The size of the pattern is of order the wavelength of the light, in fact from optics it can be shown to be $\sim \lambda f/d$ where d is the diameter of the object lens of the microscope, f the focal distance (the distance from the lens to the object).

We shall take $f/d \sim 1$, as it usually is. So looking at an object the size of an electron should give a diffraction pattern centered on the location of the object. That would seem to pin down its position fairly precisely.

What about the *momentum* of the electron? Here a problem arises that doesn't matter for larger objects the light we see has, of course, bounced off the electron, and so the electron has some recoil momentum. That is, by bouncing light off the electron we have given it some momentum.

Can we say how much? To make it simple, suppose we have good eyes and only need to bounce one photon off the electron to see it. We know the initial momentum of the photon (because we know the direction of the light beam we're using to illuminate the electron) and we know that after bouncing off, the photon hits the object lens and goes through the microscope, but we don't know *where* the photon hit the object lens. The whole point of a microscope is that all the light from a point, light that hits the object lens *in different places*, is all focused back to one spot, forming the image (apart from the blurriness mentioned above).

So if the light has wavelength λ, its constituent photons have momentum $\sim h/\lambda$, and from our ignorance of where the photon entered the microscope we are uncertain of its x-direction momentum by an amount $\sim h/\lambda$. Necessarily, then, we have the same uncertainty about the electron's x-direction momentum, since this was imparted by the photon bouncing off.

But now we have a problem. In our attempts to minimize the uncertainty in the electron's momentum, by only using one photon to detect it, we are not going to see much of the diffraction pattern discussed above such diffraction patterns are generated by *many* photons hitting the film, retina or whatever detecting equipment is being used. A single photon generates a single point (at best!). This point will most likely be within of order λ of the centre of the pattern, but this leaves us with an uncertainty in position of order λ. Therefore, in attempting to observe the position and momentum of a single electron using a single photon, we find an uncertainty in position $\Delta x \sim \lambda$, and in momentum $\Delta px \sim h/\lambda$.

These results are in accordance with Heisenberg's Uncertainty Principle $\Delta x.\Delta p_x \sim h$. Of course, we could pin down the position much better if we used N photons instead of a single one. From statistical theory, it is known that the remaining uncertainty $\sim \lambda/\sqrt{N}$. But then N photons have bounced off the electron, so, since each is equally likely to have gone through any part of the object lens, the uncertainly in momentum of the electron as a consequence of these collisions goes up as $\sqrt{N}$. (The same as the average imbalance between heads and tails in a sequence of N coin flips.)

Noting that the uncertainty in the momentum of the electron arises because we don't know where the bounced-off photon passes through the object lens, it is tempting to think we could just use a smaller object lens, that would reduce Δp_x.

Although this is correct, recall from above that we stated the size of the diffraction pattern was $\sim\lambda f/d$, where d is the diameter of the object lens and f its focal length. It is easy to see that the diffraction pattern, and consequently Δx, gets bigger by just the amount that Δp_x gets smaller!

Watching Electrons in the Double Slit Experiment

Suppose now that in the double slit experiment, we set out to detect which slit each electron goes through by shining a light just behind the screen and watching for reflected light from the electron immediately after it had passed through a slit. Following the discussion in Feynman's *Lectures in Physics,* Volume III, we shall now establish that if we can detect the electrons, we ruin the diffraction pattern!

$$\left(n+\frac{1}{2}\right)\lambda_{\text{elec}} = d \sin \theta.$$

Taking the distance between the two slits to be d, the dark lines in the diffraction pattern are at angles If the light used to see which slit the electron goes through generates an uncertainty in the electron's y momentum Δp_y, in order not to destroy the diffraction pattern we must have

$$\Delta p_y / p < \lambda_{\text{elec}} / d$$

(the angular uncertainty in the electron's direction must not be enough to spread it from the diffraction pattern maxima into the minima). Here p is the electron's full momentum, $p = h/\lambda_{\text{elec}}$. Now, the uncertainty in the electron's y momentum, looking for it with a microscope, is $\Delta p_y \sim h/\lambda_{\text{light}}$.

Substituting these values in the inequality above we find the condition for the diffraction pattern to survive is

$$\lambda_{\text{light}} > d,$$

the wavelength of the light used to detect which slit the electron went through must be greater than the distance between the slits. Unfortunately, the light scattered from the electron then gives one point in a diffraction pattern of size the wavelength of the light used, so even if we see the flash this does not pin down the electron sufficiently to say which slit it went through. Heisenberg wins again.

How the Uncertainty Principle Determines the Size of Everything

It is interesting to see how the actual physical size of the hydrogen atom is determined by the wave nature of the electron, in effect, by the Uncertainty Principle. In the ground state of the hydrogen atom, the electron minimizes its total energy. For a classical atom, the energy would be minus infinity, assuming the nucleus is a point (and very large in any case) because the electron would sit right on top of the nucleus. However, this cannot happen in quantum mechanics. Such a very localized electron would have a very large uncertainty in momentum in other words, the kinetic energy would be large. This is most clearly seen by imagining that the electron is going in a circular orbit of radius r with angular momentum $h/2\pi$. Then one wavelength of the electron's de Broglie wave just fits around the circle, $\lambda_{elec} = 2\pi r$. Clearly, as we shrink the circle's radius r, λ_{elec} goes down proportionately, and the electrons momentum,

$$p = h/\lambda_{elec} = h/2\pi r$$

increases. Adding the electron's electrostatic potential energy we find the total energy for a circular orbit of radius r is:

$$E(r) = K.E. + P.E. = \frac{p^2}{2m} - \frac{e^2}{4\pi\varepsilon_0 r} = \frac{h^2}{8m\pi^2 r^2} - \frac{e^2}{4\pi\varepsilon_0 r}.$$

Notice that for very large r, the potential energy dominates, the kinetic energy is negligible, and shrinking the atom lowers the total energy.

However, for small enough r, the (always positive) kinetic energy term wins, and the total energy *grows* as the atom shrinks. Evidently, then, there must be a value of r for which *the total energy is a minimum*. Visualizing a graph of the total energy given by the equation above as a function of r, at the minimum point the slope of $E(r)$ is zero, $dE(r)/dr = 0$.

That is,

$$-\frac{h^2}{4m\pi^2 r} + \frac{e^2}{4\pi\varepsilon_0} = 0$$

giving,

$$r_{min} = \frac{\varepsilon_0 h^2}{\pi mc^2}.$$

The total energy for this radius is the exact right answer, which is reassuring.

The point of this exercise is to see that in quantum mechanics, unlike classical mechanics, a particle cannot position itself at the exact minimum of potential energy, because that would require a very narrow wave packet and thus be expensive in kinetic energy.

The ground state of a quantum particle in an attractive potential is a trade off between potential energy minimization and kinetic energy minimization. Thus the physical sizes of atoms, molecules and ultimately ourselves are determined by Planck's constant.

Schrodinger's Equation

Photons and Electrons

We have seen that electrons and photons behave in a very similar fashion both exhibit diffraction effects, as in the double slit experiment, both have particle like or quantum behaviour. We can in fact give a complete analysis of photon behaviour we can figure out how the electromagnetic wave propagates, using Maxwell's equations, then find the probability that a photon is in a given small volume of space *dxdydz*, is proportional to $|E|^2 dxdydz$, the energy density. On the other hand, our analysis of the electron's behaviour is incomplete we know that it must also be described by a wave function $\psi(x, y, z, t)$ analogous to E, such that $|\psi(x, y, z, t)|^2\, dxdydz$ gives the probability of finding the electron in a small volume *dxdydz* around the point (x, y, z) at the time t. However, we do not yet have the analog of Maxwell's equations to tell us how ψ varies in time and space. A plausible derivation of such an equation by examining how the Maxwell wave equation works for a single-particle (photon) wave, and constructing parallel equations for particles which, unlike photons, have nonzero rest mass.

Maxwell's Wave Equation

Let us examine what Maxwell's equations tell us about the motion of the simplest type of electromagnetic wave a monochromatic wave in empty space, with no currents or charges present. First, we briefly review the derivation of the wave equation from Maxwell's equations in empty space:

$$\operatorname{div} \vec{B} = 0$$

$$\operatorname{div} \vec{E} = 0$$

$$\operatorname{curl} \vec{E} = -\frac{\partial \vec{B}}{\partial t}$$

$$\operatorname{curl} \vec{B} = \frac{1}{c^2}\frac{\partial \vec{E}}{\partial t}$$

To derive the wave equation, we take the curl of the third equation:

$$\operatorname{curl} \vec{E} = -\frac{\partial \vec{B}}{\partial t} \operatorname{curl} \vec{B} = -\frac{1}{c^2}\frac{\partial \vec{E}}{\partial t}$$

together with the vector operator identity

$$\text{curl curl} = \text{grad(div)} \times \nabla^2$$

to give,

$$\nabla^2 \vec{E} - \frac{1}{c^2}\frac{\nabla^2 \vec{E}}{\partial t^2} = 0.$$

For a plane wave moving in the x-direction this reduces to,

$$\frac{\nabla^2 \vec{E}}{\partial x^2} - \frac{1}{c^2}\frac{\nabla^2 \vec{E}}{\partial t^2} = 0$$

The monochromatic solution to this wave equation has the form:

$$\vec{E}(x,t) = \vec{E}_0 e^{i(kx-\omega t)}.$$

(Another possible solution is proportional to $\cos(kx-\omega t)$. We shall find that the exponential form, although a complex number, proves more convenient. The physical electric field can be taken to be the real part of the exponential for the classical case.) Applying the wave equation differential operator to our plane wave solution,

$$\left(\frac{\partial^2}{\partial x^2} - \frac{1}{c^2}\frac{\partial^2}{\partial t^2}\right)\vec{E}_0 e^{i(kx-\omega t)} = \left(k^2 - \frac{\omega^2}{c^2}\right)\vec{E}_0 e^{i(kx-\omega t)} = 0$$

If the plane wave is a solution to the wave equation, this must be true for all x and t, so we must have.

$$\omega = ck.$$

This is just the familiar statement that the wave must travel at c.

What does the Wave Equation Tell us about the *Photon*?

We know from the photoelectric effect and Compton scattering that the photon energy and momentum are related to the frequency and wavelength of the light by,

$$E = h\upsilon = \hbar\omega$$

$$p = \frac{h}{\lambda} = \hbar k$$

Notice, then, that the wave equation tells us that $\omega = ck$ and hence $E = cp$. To put it another way, if we think of $e^{i(kx-\omega t)}$ as describing a particle (photon) it would be more natural to write the plane wave as,

$$\vec{E}_0 e^{\frac{i}{\hbar}(px-Bt)}$$

that is, in terms of the energy and momentum of the particle. In these terms, applying the (Maxwell) wave equation operator to the plane wave yields,

$$\left(\frac{\partial^2}{\partial x^2}-\frac{1}{c^2}\frac{\partial^2}{\partial t^2}\right)\vec{E}_0 e^{\frac{i}{\hbar}(px-Bt)} = \left(p^2-\frac{E^2}{c^2}\right)\vec{E}_0 e^{\frac{i}{\hbar}(px-Bt)} = 0$$

or

$$E^2 = c^2p^2.$$

The wave equation operator applied to the plane wave describing the particle propagation yields the energy-momentum relationship for the particle.

Constructing a Wave Equation for a Particle with Mass

The discussion above suggests how we might extend the wave equation operator from the photon case (zero rest mass) to a particle having rest mass m_0. We need a wave equation operator that, when it operates on a plane wave, yields

$$E^2 = c^2p^2 + m_0^2c^4$$

Writing the plane wave function,

$$\varphi(x,t) = Ae^{\frac{i}{\hbar}(px-Bt)}$$

where A is a constant, we find we can get $E^2 = c^2p^2 + m_0^2c^4$ by adding a constant (mass) term to the differentiation terms in the wave operator:

$$\left(\frac{\partial^2}{\partial x^2}-\frac{1}{c^2}\frac{\partial^2}{\partial t^2}-\frac{m_0^2c^2}{\hbar^2}\right)Ae^{\frac{i}{\hbar}(px-Bt)} =$$

$$-\frac{1}{\hbar^2}\left(p^2-\frac{1}{c^2}\frac{E^2}{c^2}+m_0^2c^2\right)Ae^{\frac{i}{\hbar}(px-Bt)} = 0.$$

This wave equation is called the *Klein-Gordon* equation and correctly describes the propagation of relativistic particles of mass m_0. However, it's a bit inconvenient for nonrelativistic particles, like the electron in the hydrogen atom, just as, $E^2 = m_0^{\,2}c^4 + c^2p^2$ is less useful than $E= p^2/2m$ for this case.

A Nonrelativistic Wave Equation

Continuing along the same lines, let us assume that a nonrelativistic electron in free space (no potentials, so no forces) is described by a plane wave:

$$\psi(x, t) = Ae^{\frac{i}{\hbar}(px-Bt)}.$$

We need to construct a wave equation operator which, applied to this wave function, just gives us the ordinary nonrelativistic energy-momentum relationship, $E = p^2/2m$. The p^2 obviously comes as usual from differentiating twice with respect to x, but the only way we can get E is by having a *single* differentiation with respect to time, so this looks different from previous wave equations:

$$i\hbar\frac{\partial\psi(x,t)}{\partial t} = -\frac{\hbar^2}{2m}\frac{\partial^2\psi(x,t)}{\partial x^2}..$$

This isSchrodinger *'s equation* for a free particle. It is easy to check that if $\psi(x, t)$ has the plane wave form given above, the condition for it to be a solution of this wave equation is just $E = p^2/2m$. Notice one remarkable feature of the above equation the i on the left means that ψ *cannot* be a real function.

How Does a Varying Potential Affect a de Broglie Wave?

The effect of a potential on a de Broglie wave was considered by Sommerfeld in an attempt to generalize the rather restrictive conditions in Bohr's model of the atom.

Since the electron was orbiting in an inverse square force, just like the planets around the sun, Sommerfeld couldn't understand why Bohr's atom had only circular orbits, no Kepler-like ellipses.

De Broglie's analysis of the allowed circular orbits can be formulated by assuming that at some instant in time the spatial variation of the wave function on going around the orbit includes a phase term of the form $e^{\frac{i}{\hbar}pq}$, where here the parameter q measures distance around the orbit. Now for an acceptable wave function, the total phase change on going around the orbit must be $2n\pi$, where n is an integer. For the usual Bohr circular orbit, p is constant on going around, q changes by $2\pi r$, where r is the radius of the orbit, giving

$$\frac{1}{\hbar}p2\pi r = 2n\pi\ \psi$$

so,

$$pr = n\hbar,$$

the usual angular momentum quantization. What Sommerfeld did was to consider a general Kepler ellipse orbit, and visualize the wave going around such an orbit. Assuming the usual relationship $p = h/$ ë, the wavelength will vary as the particle moves around the orbit, being shortest where the particle moves fastest, at its closest approach to the nucleus. Nevertheless, the phase change on moving a short distance Δq should still be,

$$\frac{1}{\hbar}p\Delta q\,,$$

and requiring the wave function to link up smoothly on going once around the orbit gives,

$$\int pdq = nh$$

Thus only certain elliptical orbits are allowed. The mathematics is nontrivial, but it turns out that every allowed elliptical orbit has the same energy as one of the allowed circular orbits. This is why Bohr's theory gave all the energy levels. Actually, this whole analysis is old fashioned (it's called the "old quantum theory") but we've gone over it to introduce the idea of *a wave with variable wavelength, changing with the momentum as the particle moves through a varying potential.*

Schrodinger's Equation for a Particle in a Potential

Let us consider first the one-dimensional situation of a particle going in the x-direction subject to a "roller coaster" potential. What do we expect the wave function to look like? We would expect the wavelength to be shortest where the potential is lowest, in the valleys, because that's where the particle is going fastest—maximum momentum. Perhaps slightly less obvious is that the amplitude of the wave would be largest at the tops of the hills (provided the particle has enough energy to get there) because that's where the particle is moving slowest, and therefore is most likely to be found. With a nonzero potential present, the energy-momentum relationship for the particle becomes the energy equation,

$$E = \frac{p^2}{2m} + V(x)\,.$$

We need to construct a wave equation which leads naturally to this relationship. In contrast to the free particle cases discussed above, the relevant wave function here will no longer be a plane wave, since the wavelength varies with the potential. However, at a given x, the momentum is determined by the "local wavelength", that is,

$$p = -i\hbar \frac{\partial \psi}{\partial x}.$$

It follows that the appropriate wave equation is:

$$i\hbar \frac{\partial \psi(x,t)}{\partial t} = \frac{\hbar^2}{2m} \frac{\partial^2 \psi(x,t)}{\partial x^2} + V(x)\psi(x,t).$$

This is the standard one-dimensionalSchrodinger equation. In three dimensions, the argument is precisely analogous. The only difference is that the square of the momentum is now a sum of three squared components, for the x, y and z directions, so,

$$\frac{\partial^2}{\partial x^2}$$

becomes,

$$\frac{\partial^2}{\partial x^2} + \frac{\partial^2}{\partial y^2} + \frac{\partial^2}{\partial z^2} = \nabla^2,$$

and the equation is:

$$i\hbar \frac{\partial \psi(x,y,z,t)}{\partial t} = \frac{\hbar^2}{2m} \nabla^2 \psi(x,y,z,t) + V(x,y,z)\psi(x,y,z,t).$$

This is the completeSchrodinger equation.

Electron in a Box

Plane Wave Solutions

The best way to gain understanding ofSchrodinger 's equation is to solve it for various potentials. The simplest is a one-dimensional "particle in a box" problem. The appropriate potential is $V(x) = 0$ for x between 0, L and $V(x)$ = infinity otherwise—that is to say, there are infinitely high walls at $x = 0$ and $x = L$, and the particle is trapped between them. This turns out to be quite a good approximation for electrons in a long molecule, and the three-dimensional version is a reasonable picture for electrons in metals. Between $x = 0$ and $x = L$ we have $V = 0$, so the wave equation is just

$$i\hbar \frac{\partial \psi(x,t)}{\partial t} = -\frac{\hbar^2}{2m} \frac{\partial^2 \psi(x,t)}{\partial x^2}$$

A possible plane wave solution is,

$$(x,\ t) = Ae^{\frac{i}{\hbar}(px - Bt)}.$$

On inserting this into the zero-potentialSchrodinger equation above we find $E = p^2/2m$, as we expect. It is very important to notice that the complex conjugate, proportional to,

$$e^{-\frac{i}{\hbar}(px-Bt)},$$

is *not* a solution to theSchrodinger equation! If we blindly put it into the equation we get

$$E = -p^2/2m,$$

an unphysical result. However, a wave function proportional to,

$$e^{-\frac{i}{\hbar}(px+Bt)}$$

gives $E = p^2/2m$, so this plane wave *is* a solution to the equation. Therefore, the two allowed plane-wave solutions to the zero-potentialSchrodinger equation are proportional to,

$$e^{-\frac{i}{\hbar}(px-Bt)} \text{ and,}$$

$$e^{-\frac{i}{\hbar}(px+Bt)}$$

respectively. Note that these two solutions have the *same time dependence,*

$$e^{-\frac{iBt}{\hbar}}.$$

To decide on the appropriate solution for our problem of an electron in a box, of course we have to bring in the walls—what they mean is that $\psi = 0$ for $x < 0$ and for $x > L$ because remember $|\psi|^2$ tells us the probability of finding the particle anywhere, and, since it's in the box, it's trapped *between* the walls, so there's zero probability of finding it outside. The condition $\psi = 0$ at $x = 0$ and $x = L$ reminds us of the vibrating string with two fixed ends—the solution of the string wave equation is standing waves of sine form. In fact, taking the difference of the two permitted plane-wave forms above gives a solution of this type:

$$\psi(x, t) = A\sin\frac{px}{\hbar}e^{-\frac{iBt}{\hbar}}.$$

This wave function satisfies the Schrodinger equation between the walls, it vanishes at the $x = 0$ wall, it will also vanish at $x = L$ provided that the momentum variable satisfies:

$$\frac{pL}{\hbar} = \pi,\ 2\pi,\ 3\pi...$$

Thus the allowed values of p are $hn/2L$, where $n = 1, 2, 3...$, and from $E = p^2/2m$ the allowed energy levels of the particle are:

$$E = \frac{p^2}{2m} = \frac{1}{2m}\left(\frac{h}{2L}\right)^2, \frac{4}{2m}\left(\frac{h}{2L}\right)^2, \frac{9}{2m}\left(\frac{h}{2L}\right)^2, ...$$

Note that these energy levels become more and more widely spaced out at high energies, in contrast to the hydrogen atom potential. (The harmonic oscillator potential gives equally spaced energy levels, so by studying how the spacing of energy levels varies with energy, we can learn something about the shape of the potential.) What about the overall multiplicative constant A in the wave function? This can be real or complex. To find its value, note that at a fixed time, say $t = 0$, the probability of the electron being between x and $x + dx$ is $|\psi|^2 dx$ or,

$$|A|^2 \sin^2 \frac{px}{\hbar} dx\ .$$

The total probability of the particle being somewhere between 0, L must be unity:

$$\int_{x-0}^{x-L} |A|^2 \sin^2 \frac{px}{\hbar} dx = 1, \quad \text{so} \frac{1}{2} L |A|^2 = 1.$$

Hence,

$$\psi(x,\ t) = \sqrt{\frac{2}{L}} \sin \frac{px}{\hbar} e^{-\frac{iBt}{\hbar}}.$$

When A is fixed in this way, by demanding that the total probability of finding the particle somewhere be unity, it is called the normalization constant.

Stationary States

Notice that at a later time the probability distribution for the wave function,

$$\psi(x,\ t) = \sqrt{\frac{2}{L}} \sin \frac{px}{\hbar} e^{-\frac{iBt}{\hbar}}.$$

is the same, because time only appears as a phase factor in this time-dependent function, and so does not affect $|\psi|^2$. A state with a time-independent probability distribution is called a stationary state.

States with Moving Probability Distributions

Recall that the Schrodinger equation is a linear equation, and the sum of any two solutions is also a solution to the equation. That means that we can add two solutions having different energies, and still have a legal wave function. We shall establish that in this case, the probability distribution varies in time.

The simplest way to see how this must be is to look at an example. Let's add the ground state to the first excited state, and normalize the sum:

$$\psi(x,\ t) = \sqrt{\frac{1}{L}}\left(\sin\frac{\pi x}{L}e^{-\frac{izkt}{4mL^2}} + \sin\frac{2\pi x}{L}e^{-\frac{izkt}{mL^2}}\right)$$

note: h,not $\hbar$. (You can check the normalization constant at $t = 0$). For general x, the two terms in the bracket rotate in the complex plane at different rates, so their sum has a time-varying magnitude. That is to say, $|\ \psi(x,t)\,|^2$ varies in time, so the particle must be moving around this is *not* a stationary state.

Exercise: To see this, note that at t = 0 the wave function is:

$$\psi(x,\ 0) = \sqrt{\frac{1}{L}}\left(\sin\frac{\pi x}{L} + \sin\frac{2\pi x}{L}\right)$$

and sketch this function: the particle is more likely to be found in the left-hand half of the box.

Now, suppose the time is,

$$t = 4mL^2/h,\ \text{so}\ e^{-\frac{izkt}{4mL^2}} = -1.$$

At this time,

$$\psi(4mL^2/h) = \sqrt{\frac{1}{L}}\left(-\sin\frac{\pi x}{L} + \sin\frac{2\pi x}{L}\right)$$

and it's easy to see that the particle is more likely to be found n the *right*-hand half. That is to say, this wave function, a linear sum of wave functions corresponding to different energies, has a probability distribution that sloshes back and forth in the box: and, any attempt to describe a classical-type particle motion, bouncing back and forth, *necessarily* involves adding quantum wave functions of different energies.

Note that the frequency of the sloshing motion depends on the *difference* of the two energies: how constructively the two components interfere depends on the difference of the phases in the energies at the time. A single energy wave function always has a static probability distribution.

Of course, the *total* probability of finding the particle *somewhere* in the box remains unity: the normalization constant is time-independent.

Eigenstates and Eigenvalues

The only way to prevent $|\psi(x,t)|^2$ varying in time is to have all its parts changing phase in time at the same rate. *This means they all correspond to the same energy.* If we restrict our considerations to such *stationary states*, the wave function can be factorized,

$$\psi(x,\ t) = \psi(x)\, e^{\frac{-iBt}{\hbar}}$$

and putting this wave function into theSchrodinger equation we find,

$$-\frac{\hbar^2}{2m}\frac{d^2\psi(x)}{dx^2} + V(x)\psi(x) = E\psi(x).$$

This is the time-independentSchrodinger equation, and its solutions are the spatial wave functions for stationary states, states of definite energy. These are often called eigenstates of the equation. The values of energy corresponding to these eigenstates are called the eigenvalues.

Wavefunction for the Lowest Energy

Consider again the wavefunction for the lowest energy state of a particle confined between walls at $x = 0$ and $x = L$. The reader should sketch the wavefunction from some point to the left of $x = 0$ over to the right of $x = L$.

To the left of $x = 0$, the wavefunction is exactly zero, then at $x = 0$ it takes off to the right (inside the box) as a sine curve. In other words, at the origin the slope of the wavefunction ψ is zero to the left, nonzero to the right. There is a discontinuity in the slope at the origin: this means the second derivative of ψ is infinite at the origin.

On examining the time-independent Schrodinger equation above, we see the equation can *only* be satisfied at the origin because the potential becomes infinite there—the wall is an infinite potential.

(And, in fact, since ψ becomes zero on approaching the origin from inside the box, the limit must be treated carefully.)

It now becomes obvious that if the box does not have infinite walls, but merely high ones, ψ describing a confined particle cannot suddenly go to zero at the walls: the second derivative must remain finite. For non-infinite walls, ψ and its derivative must be continuous on entering the wall.

This has the important physical consequence that ψ will be nonzero at least for some distance into the wall, even if classically the confined particle does not have enough energy to "climb the wall". (Which it doesn't, if it's confined.) Thus, in quantum mechanics, there is a non-vanishing probability of finding the particle in a region which is "classically forbidden" in the sense that it doesn't have enough energy to get there.

Rutherford Scattering

In 1908 Rutherford was awarded the Nobel Prize-for chemistry! The award citation read: "for his investigations into the disintegration of the elements, and the chemistry of radioactive substances."

While at McGill University, he had discovered that the radioactive element thorium emitted a gas which was itself radioactive, but if the gas radioactivity was monitored separately from the thorium's, he found it decreased geometrically, losing approximately half its current strength for each minute that passed. The gas he had found was a short-lived isotope of radon, and this was the first determination of a "half-life" for a radioactive material.

The chemists were of course impressed that Rutherford was fulfilling their ancient alchemical dream of transmuting elements, or at least demonstrating that it happened. Rutherford himself remarked at the ceremony that he "had dealt with many different transformations with various time-periods, but the quickest he had met was his own transformation from a physicist to a chemist".

Still, Nobel prizes of any kind are nice to get, so he played along, titling his official Nobel lecture: "The chemical nature of the alpha-particle from radioactive substances". (He established that his favourite particle was an ionized helium atom by collecting alphas in an evacuated container, where they picked up electrons. After compressing this very rarefied gas, he passed an electric discharge through it and observed the characteristic helium spectrum in the light emitted.)

Rutherford was the world leader in alpha-particle physics. In 1906, at McGill, he had been the first to detect slight deflections of alphas on passage through matter. In 1907, he became a professor at the University of Manchester, where he worked with Hans Geiger.

This was just a year after Rutherford's old boss, J. J. Thomson, had written a paper on his plum pudding atomic model suggesting that the number of electrons in an atom was about the same as the atomic number. (Not long before, people had speculated that atoms might contain thousands of electrons.

They were assuming that the electrons contributed a good fraction of the atom's mass.) The actual distribution of the electrons in the atom, though, was as mysterious as ever. Mayer's magnets were fascinating, but had not led to any quantitative conclusions on electronic distributions in atoms. Rutherford's 1906 discovery that his pet particles were slightly deflected on passing through atoms came about when he was finding their charge to mass ratio, by measuring the deflection in a magnetic field.

He detected the alphas by letting them impact photographic film. When he had them pass through a thin sheet of mica before hitting the film (so the film didn't have to be in the vacuum?) he found the image was blurred at the edges, evidently the mica was deflecting the alphas through a degree or two.

He also knew that the alphas wouldn't be deflected a detectable amount by the *electrons* in the atom, since the alphas weighed 8,000 times as much as the electrons, atoms contained only a few dozen electrons, and the alphas were very fast. The mass of the atom must be tied up somehow with the positive charge. Therefore, he reasoned, analyzing these small deflections might give some clue as to the distribution of positive charge and mass in the atom, and therefore give some insight into his old boss J. J.'s plum pudding. The electric fields necessary in the atom for the observed scattering already seemed surprisingly high to Rutherford.

Photoelectric Effect

The most dramatic prediction of Maxwell's theory of electromagnetism, published in 1865, was the existence of electromagnetic waves moving at the speed of light, and the conclusion that light itself was just such a wave. This challenged experimentalists to generate and detect electromagnetic radiation using some form of electrical apparatus. The first clearly successful attempt was by Heinrich Hertz in 1886. He used a high voltage induction coil to cause a spark discharge between two pieces of brass, to quote him, "Imagine a cylindrical brass body, 3 cm in diameter and 26 cm long, interrupted midway along its length by a spark gap whose poles on either side are formed by spheres of 2 cm radius." The idea was that once a spark formed a conducting path between the two brass conductors, charge would rapidly oscillate back and forth, emitting electromagnetic radiation of a wavelength similar to the size of the conductors themselves.

To prove there really was radiation emitted, it had to be detected. Hertz used a piece of copper wire 1 mm thick bent into a circle of diameter 7.5 cms, with a small brass sphere on one end, and the other end of the wire was pointed, with the point near the sphere. He added a screw mechanism so that the point could be moved very close to the sphere in a controlled fashion. This "receiver" was designed so that current oscillating back and forth in the wire would have a natural period close to that of the "transmitter" described above. The presence of oscillating charge in the receiver would be signaled by a spark across the (tiny) gap between the point and the sphere (typically, this gap was hundredths of a millimeter). (It was suggested to Hertz that this spark gap could be replaced as a detector by a suitably prepared frog's leg, but that apparently didn't work.)

The experiment was very successful-Hertz was able to detect the radiation up to fifty feet away, and in a series of ingenious experiments established that the radiation was reflected and refracted as expected, and that it was polarized. The main problem-the limiting factor in detectionwas being able to see the tiny spark in the receiver. In trying to improve the spark's visibility, he came upon something very mysterious.

To quote from Hertz again (he called the transmitter spark A, the receiver B): "I occasionally enclosed the spark B in a dark case so as to more easily make the observations; and in so doing I observed that the maximum spark-length became decidedly smaller in the case than it was before. On removing in succession the various parts of the case, it was seen that the only portion of it which exercised this prejudicial effect was that which screened the spark B from the spark A.

The partition on that side exhibited this effect, not only when it was in the immediate neighbourhood of the spark B, but also when it was interposed at greater distances from B between A and B. A phenomenon so remarkable called for closer investigation."

Hertz then embarked on a very thorough investigation. He found that the small receiver spark was more vigourous if it was exposed to ultraviolet light from the transmitter spark. It took a long time to figure this out-he first checked for some kind of electromagnetic effect, but found a sheet of glass effectively shielded the spark.He then found a slab of quartz did not shield the spark, whereupon he used a quartz prism to break up the light from the big spark into its components, and discovered that the wavelength which made the little spark more powerful was beyond the visible, in the ultraviolet.

In 1887, Hertz concluded what must have been months of investigation: " I confine myself at present to communicating the results obtained, without attempting any theory respecting the manner in which the observed phenomena are brought about."

Hallwachs' Simpler Approach

The next year, 1888, another German physicist, Wilhelm Hallwachs, in Dresden, wrote: "In a recent publication Hertz has described investigations on the dependence of the maximum length of an induction spark on the radiation received by it from another induction spark.

He proved that the phenomenon observed is an action of the ultraviolet light. No further light on the nature of the phenomenon could be obtained, because of the complicated conditions of the research in which it appeared.

I have endeavored to obtain related phenomena which would occur under simpler conditions, in order to make the explanation of the phenomena easier. Success was obtained by investigating the action of the electric light on electrically charged bodies."

He then describes his very simple experiment: a clean circular plate of zinc was mounted on an insulating stand and attached by a wire to a gold leaf electroscope, which was then charged negatively.

The electroscope lost its charge very slowly. However, if the zinc plate was exposed to ultraviolet light from an arc lamp, or from burning magnesium, charge leaked away quickly. If the plate was positively charged, there was no fast charge leakage. Questions for the reader: Could it be that the ultraviolet light somehow spoiled the insulating properties of the stand the zinc plate was on? Could it be that electric or magnetic effects from the large current in the arc lamp somehow caused the charge leakage? Although Hallwach's experiment certainly clarified the situation, he did not offer any theory of what was going on.

J.J. Thomson Identifies the Particles

In fact, the situation remained unclear until 1899, when Thomson established that the ultraviolet light caused *electrons* to be emitted, the same particles found in cathode rays. His method was to enclose the metallic surface to be exposed to radiation in a vacuum tube, in other words to make it the cathode in a cathode ray tube.

The new feature was that electrons were to be ejected from the cathode by the radiation, rather than by the strong electric field used previously. By this time, there was a plausible picture of what was going on. Atoms in the cathode contained electrons, which were shaken and caused to vibrate by the oscillating electric field of the incident radiation. Eventually some of them would be shaken loose, and would be ejected from the cathode.

It is worthwhile considering carefully how the *number* and *speed* of electrons emitted would be expected to vary with the *intensity* and *colour* of the incident radiation. Increasing the intensity of radiation would shake the electrons more violently, so one would expect more to be emitted, and they would shoot out at greater speed, on average.

Increasing the frequency of the radiation would shake the electrons faster, so might cause the electrons to come out faster. For very dim light, it would take some time for an electron to work up to a sufficient amplitude of vibration to shake loose.

Lenard Finds Some Surprises

In 1902, Lenard studied how the energy of the emitted photoelectrons varied with the intensity of the light. He used a carbon arc light, and could increase the intensity a thousand-fold. The ejected electrons hit another metal plate, the collector, which was connected to the cathode by a wire with a sensitive ammeter, to measure the current produced by the illumination.

To measure the energy of the ejected electrons, Lenard charged the collector plate negatively, to repel the electrons coming towards it. Thus, only electrons ejected with enough kinetic energy to get up this potential hill would contribute to the current. Lenard discovered that there was a well defined minimum voltage that stopped any electrons getting through, we'll call it V_{stop}.

To his surprise, he found that V_{stop} did not depend at all on the intensity of the light! Doubling the light intensity doubled the *number* of electrons emitted, but did not affect the *energies* of the emitted electrons. The more powerful oscillating field ejected more electrons, but the maximum individual energy of the ejected electrons was the same as for the weaker field.

But Lenard did something else. With his very powerful arc lamp, there was sufficient intensity to separate out the colours and check the photoelectric effect using light of different colours. He found that the maximum energy of the ejected electrons *did* depend on the colour the shorter wavelength, higher frequency light caused electrons to be ejected with more energy.

This was, however, a fairly qualitative conclusionthe energy measurements were not very reproducible, because they were extremely sensitive to the condition of the surface, in particular its state of partial oxidation. In the best vacua available at that time, significant oxidation of a fresh surface took place in tens of minutes. The details of the surface are crucial because the fastest electrons emitted are those from right at the surface, and their binding to the solid depends strongly on the nature of the surfaceis it pure metal or a mixture of metal and oxygen atoms.

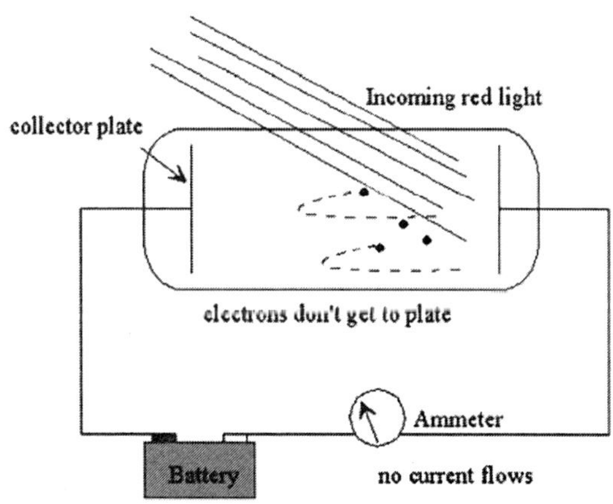

In the above figure, the battery represents the potential Lenard used to charge the collector plate negatively, which would actually be a variable voltage source. Since the electrons ejected by the blue light are getting to the collector plate, evidently the potential supplied by the battery is less than V_{stop} for blue light. Show with an arrow on the wire the direction of the electric current in the wire.

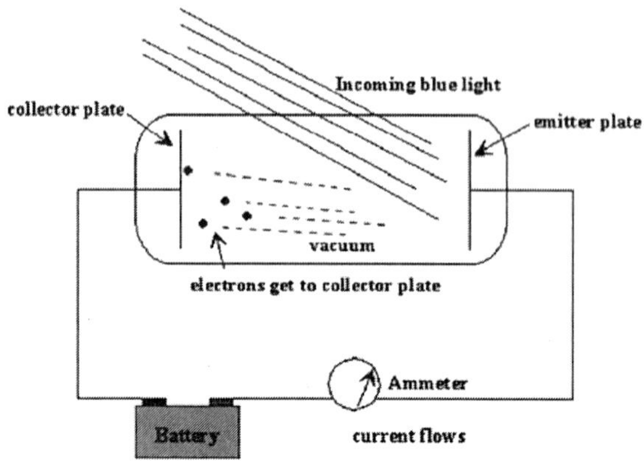

Einstein Suggests an Explanation

In 1905 Einstein gave a very simple interpretation of Lenard's results. He just assumed that the incoming radiation should be thought of as quanta of frequency hf, with f the frequency. In photoemission, one such quantum is absorbed by one electron. If the electron is some distance into the material of the cathode, some energy will be lost as it moves towards the surface. There will always be some electrostatic cost as the electron leaves the surface, this is usually called the work function, W. The most energetic electrons emitted will be those very close to the surface, and they will leave the cathode with kinetic energy,

$$E = hf - W.$$

On cranking up the negative voltage on the collector plate until the current just stops, that is, to V_{stop}, the highest kinetic energy electrons must have had energy eV_{stop} on leaving the cathode.

Thus,

$$eV_{stop} = hf - W$$

Thus Einstein's theory makes a very definite quantitative prediction: if the frequency of the incident light is varied, and V_{stop} plotted as a function of frequency, the slope of the line should be h/e. It is also clear that there is a minimum light frequency for a given metal, that for which the quantum of energy is equal to the work function. Light below that frequency, no matter how bright, will not cause photoemission.

Millikan's Attempts to Disprove Einstein's Theory

If we accept Einstein's theory, then, this is a completely different way to measure Planck's constant. The American experimental physicist Robert Millikan, who did not accept Einstein's theory, which he saw as an attack on the wave theory of light, worked for ten years, until 1916, on the photoelectric effect. He even devised techniques for scraping clean the metal surfaces inside the vacuum tube.

For all his efforts he found disappointing results: he confirmed Einstein's theory, measuring Planck's constant to within 0.5% by this method. One consolation was that he did get a Nobel prize for this series of experiments.

Sparks in Gases

Electric discharges were passed through rarefied gases in the time of Benjamin Franklin-His friend William Watson, an Englishman, remarked: "It was a most delightful spectacle, when the room was darkened, to see the electricity in its passage." The experimental arrangement was to have two small metal plates inside a glass container from which the air was gradually removed by a pump.

The plate connected to the negative side of the electricity supply was called the cathode, that to the positive side the anode. As the pressure was lowered, a spark which fattened into a purplish glowing, writhing snake appeared going from the cathode to the anode. Lowering the pressure further, as Faraday noticed in the 1830's, a dark space opened up near the cathode, now called the Faraday Dark Space.

Cathode Rays

The big breakthrough came in the 1850's when Geissler invented a much better (mercury) pump. In 1869, Hittorf found that in a very good vacuum (0.01 mm Hg), the Faraday dark space expanded to fill the whole tube, and the cathode emitted rays that caused the glass to glow where they hit. In 1876, Goldstein called them "cathode rays". In 1879, an Englishman, William Crookes, declared that they must be particles of some sort, and demonstrated that they traveled in straight lines by inserting a Maltese cross in the tube, which cast a sharp shadow on the end of the tube, a demo still in common use 120 years later! Were these cathode rays waves or particles? Hertz, with his student Lenard, discovered in 1891 that the rays could penetrate a thin aluminum plate, and detected them in the air just outside the tube. Hertz found experimentally that he could not deflect the rays with an electrostatic field applied from outside the tube, and they didn't affect a compass, so he concluded (wrongly) that they must be waves, not particles.

Thomson Discovers the Electron

For some odd reason, all the Germans physicists thought the cathode rays were waves, the English thought they were particles (presumably ions of some kind). In 1897, J. J. Thomson announced at the Royal Institution that they were negatively charged particles, and he'd measured e/m, the ratio of charge to mass for a single particle. He achieved this with a specially designed cathode ray tube, the precursor of the tube in a present-day TV set. The cathode was at the far narrow end of the tube, the anode a few centimeters away, with a small hole in its centre.

The rays went from the cathode to the anode, but a narrow beam continued through the hole in the anode to the glass at the other end of the tube, where their impact caused the glass to glow in a small central spot. Thomson also added *inside the tube* two parallel plates arranged so that the narrow beam of cathode rays went between them on its way to the end of the tube. He found that electrically charging these plates caused the beam to deflect, so the glowing spot moved from the centre of the end of the tube. (This was different from Hertz' result.

Possibly Hertz' externally applied field was shielded by space charge on or near the walls of the tube.) Thomson also added magnets to give a uniform magnetic field in the region between the plates. By adjusting this magnetic field strength to cancel the deflection of the rays caused by the electric field, he was able to measure the speed of the rays, because balancing these two forces,

$$eE = evB$$

the charge cancels, and the speed is simply the ratio of electric field strength to magnetic field strength.

Having determined the horizontal speed, Thomson was able to deduce the charge to mass ratio by measuring the deflection in the electric field alone. In this situation, the trajectory of the particle is identical to that of a ball, initially thrown horizontally, veering downwards in a parabolic path. The downward (or upward, depending on the sign of the plates' charge) force is eE, and it operates for a time L/v, where L is the length of the plates. On emerging from the region between the plates, the electron moves in a straight line.

The value of e/m that emerged from this experiment was a complete surprise. It was about 2,000 times the value for the hydrogen ion, which has the largest value of any ion. Once it had been established that the cathode rays were not just uncharged electromagnetic waves, it had been assumed that they were "molecular torrents"-a flood of the same kinds of ions found in electrolysis. Now, apparently, the cathode rays were particles smaller than the smallest atom! This was the first hint that atoms might be made up of smaller entities.

Another important point is that Thomson found this ratio e/m to be independent of the material the cathode was made of. The cathode rays could have been fragments of atoms which were different for different atoms, but evidently this was not the case. Furthermore, he found in 1899 that photoelectrically produced particles had the same e/m, so were probably the same particles.

For the photoelectrically produced particles, Thomson was able to find the charge *e*, working with his student C. T. R. Wilson, in a cloud chamber experiment. They introduced photoelectrons into a supersaturated vapour. Tiny droplets formed around each free electron. Thomson and Wilson estimated the total number of droplets formed by counting, then found the total charge by collecting the droplets on an electroscope. Their results were off by almost 40%, but close enough to suggest strongly that the particles had the same magnitude charge as the hydrogen ion. This confirmed that the charged particles were much lighter than atoms. (They could have been the same mass as atoms, but carrying a huge charge. This unlikely possibility was now ruled out.)

Thomson had discovered the electron. (Actually, the word electron was first used in 1891 by Stoney, an Irish physicist, to denote the charge on an ion, which he estimated from electrolysis and Avogadro's number.) The emerging picture was that the atom, known of course to be electrically neutral, contained negatively charged particles, the electrons, which could be removed in various ways, leaving a positively charged ion which contained almost all the mass of the original atom.

To put it in his own words, in 1899: "Electrification essentially involves the splitting up of the atom, a part of the mass of the atom getting free and becoming detached from the original atom."(Pais p 86). The word atom comes from the Greek, and means that which cannot be cut up. ("tom" is the same root for cut that appears, for example, in appendectomy). Thomson was the first to talk about splitting atoms.

It had long been known, of course, that in electrolysis of water, for example, by passing an electric current through it, the molecules were somehow split into hydrogen ions, which carried a positive charge and moved to the cathode, where they received a negative electric charge from the battery and appeared as hydrogen atoms combined into hydrogen molecules; and negatively charged oxygen ions, appearing in the same way as oxygen gas at the anode. The new insight was that these ions were atoms having an excess or a deficiency of electrons, and the electrons were particles whose charge and mass was now known.

X-rays

But actually in recounting this tale we skipped over an even more important related development in 1895. (Pais p 36) Wilhelm Roentgen was a 50 year old Director of a Physics Institute, a position that included generous family living quarters, including for example, a well stocked wine cellar.

He was checking out the work of Hertz and Lenard on cathode rays, and in particular was interested in Lenard's discovery that the rays penetrated a little way into the air if the Crookes' tube had an aluminum window.

He had covered his Crookes' tube with black cardboard, presumably (my guess) so that he could detect the rays in the air without the distracting glow from the tube itself. He darkened the room to check that no light was getting through, and suddenly (from an unnamed biographer quoted in the Project Physics Text) "noticed a weak light shimmering on a little bench nearby. Highly excited, Roentgen lit a match, and, to his great surprise, discovered that the source of the mysterious light was a little barium platinocyanide screen lying on the bench."

This little screen had been made to detect ultraviolet light, which causes barium platinocyanide to fluoresce. But if the Crookes' tube was producing any ultraviolet light, it wouldn't make it through the cardboard, and the cathode rays themselves only traveled a few centimeters in air. The little screen was a meter away from the tube. Roentgen concluded that the Crookes' tube must be emitting some new radiation of an unknown type. To reflect his bewilderment, he dubbed the new radiation "x-rays". He couldn't believe it. He checked and rechecked. He told his wife that when he announced it, everybody would say "Roentgen has probably gone crazy".

Before publishing his results, he spent seven weeks investigating the properties of these x-rays. He had already found that ordinary film was sensitive to them. He found lead was a good shield, and, by using it to shield different parts of the Crookes' tube, established that the x-rays originated in the part of the glass that fluoresced where the cathode rays hit it.

He found that the x-rays traveled in straight lines, and (unlike the cathode rays) were not deflected by magnetic fields. He found they passed through flesh almost unimpeded, but bone cast a shadow. By having his wife place her hand between the point source of x-rays on the Crookes' tube and some unexposed film in a box, then developing the film, he took a picture of the bones in her hand. On January 1, 1896 he announced his findings, complete with the bone picture, to many of his colleagues, and remarked to his wife "Now the fun begins". The x-ray story is a good antidote to the myth that new scientific discoveries are developed and utilized much more rapidly than they used to be. On the 25th of January, a young research student in Cambridge, Ernest Rutherford, wrote of his Professor: "The Professor, of course, is trying to find out the real cause and nature of the waves, and the great object is to find the theory of matter before anyone else, for nearly every Professor in Europe is now on the warpath."

Meanwhile, in the USA, the New York Times reported on the 9th of February: "the wizard of New Jersey (Edison) will try to photograph the skeleton of a human head next week." (He didn't succeed.) Already in 1896 several hospitals had x-ray facilities, and x-ray photographs were ruled as acceptable evidence in courts in France, England and the USA.

Before the end of the year, the dangers became apparent. A Columbia professor who had been giving demonstrations at Bloomingdale's suffered severe skin damage. Over the next five years, many doctors and patients suffered horribly, and patients began to sue successfully. By 1903, lead-impregnated rubber shielding devices were being used.

What are X-rays?

All the activity described above did little to clarify the actual nature of x-rays. Roentgen himself found they were undeflected by a magnetic field, so were not charged particles like the cathode rays. On the other hand, they didn't exhibit any diffraction phenomena, so did not seem to be waves. Roentgen, and independently Thomson, found that the x-rays were ionizing radiation-as they passed through air, ions were created. A gold leaf electroscope exposed to the x-rays would lose its charge, as the newly created ions were attracted to the charged leaves. In 1899, Haga and Wind noticed a slight broadening of an x-ray beam after it passed through a slit a few thousandths of a millimeter wide. This could be from diffraction if the wavelength were of order 10^{-10} meters. This problem was not resolved conclusively until 1912, when Laue made the observation that since the wavelength of x-rays was apparently similar to the distances between planes of atoms in a crystal, perhaps a crystal would act as a diffraction grating for x-rays. This turned out to be correct, and in fact is now the standard way of finding crystal structure. It was concluded, then, that x-rays were "ultra- ultra violet light", as one French physicist had put it in 1896. Once the usefulness of x-rays was established, techniques for producing them evolved rapidly. It was found that they were produced far more copiously if the cathode rays impinged on a piece of heavy metal, such as Molybdenum, rather than glass. The physical picture of x-ray production was that the electrons radiated as they suddenly decelerated on hitting the target, unloading their kinetic energy as radiation (plus some heating of the target). This deceleration radiation is called bremsstrahlung in German, and this word is sometimes used to describe it.

X-rays come in quanta

Since crystals act as diffraction gratings for x-rays, it is possible to map out a spectrum of the x-rays. (Of course, this was not done until after 1912). The most important finding on doing this was that there was a maximum frequency (minimum wavelength) at which x-ray production stopped, and this maximum frequency depended on the voltage applied to the cathode ray tube in a simple way:

$$hf_{max} = eV$$

The explanation is of course very simple-each electron in the cathode rays gains a kinetic energy of eV on accelerating down the tube before crashing into the target. Therefore, this is the most energy that can be emitted as it loses kinetic energy in the crash. If this all goes into one photon, the frequency will be f_{max}. So the cutoff in the x-ray spectrum provides another way to measure Planck's constant.

More Rays

To return now to 1896, at first nobody had any idea how the x-rays were being generated, but Roentgen had clearly established that they came from the bit of Crookes' tube glass that was fluorescing. A French physicist, Henri Becquerel, attended a meeting of the French

Academy of Sciences on January 20, 1896, *three weeks* after Roentgen's first announcement. Two French physicians were already showing hand x-rays. It occurred to Becquerel that any substance that fluoresced intensely might also emit x-rays. It happened that fifteen years previously he had worked with a substance that fluoresced strongly when exposed to sunlight, so he wondered if it might also be emitting x-rays, previously undetected.

To find out, he placed a sheet of it in the sun, lying on top of a cardboard box containing unexposed film with a small metal object above it. After a day's exposure to the sunlight, which caused fluorescence, he took out and developed the film. Sure enough, it was exposed except in the shadow of the metal object, which appeared in silhouette. He decided to check, and set up the same arrangement again, with a small metal cross above the film. Unfortunately, however, it continued cloudy, so he kept the package in a closet waiting for the sun to reappear. After several days Becquerel grew impatient, and, possibly not having enough to do, decided to develop the film, perhaps to check if some light had leaked in. To his utter astonishment, he found the silhouette of the cross on the film just as distinctly as before! *Question for the reader*: before reading on, can you figure out any possible explanation for this surprising discovery? Evidently, the recent exposure to sunlight had not been a crucial element in the production of the mysterious radiation that had exposed the film. Becquerel theorized that maybe the substance continued to manufacture rays long after the fluorescence died away. However, it soon became clear that in fact the fluorescence itself was irrelevant.

It happened that the fluorescent substance Becquerel used for his experiment contained uranium. After extensive experimenting with various substances, he concluded by May that the radiation came from any substances containing uranium, whether they fluoresced or not. He found the intensity of the radiation increased with the amount of uranium present, and did not appear to change in intensity with time, or temperature, or chemical action. Like x-rays, the radiation was ionizing. If a piece of uranium is held near an electroscope, the electroscope discharges. Unlike x-rays, though, these rays couldn't be turned off.

Marie Curie Investigates Becquerel's Rays

In contrast to all the hoopla over x-rays, Becquerel's rays had little public impact, at least at the beginning. Fortunately, they did attract the attention of two very talented young researchers, Marie Curie in Paris and Ernest Rutherford in Cambridge. Marie worked with her husband Pierre, carefully measuring the amount of radiation from different amounts of uranium compounds. They measured the radiation by using two parallel metal plates differing in potential by 100 volts, connected by a sensitive electrometer (basically an infinite resistance voltmeter, a calibrated electroscope).

A thin layer of the uranium compound under examination was then put on the top plate, and the rate at which charge was lost was measured-the ionizing radiation caused the charge leakage. Marie concluded that the radiation intensity was just proportional to the number of uranium atoms present, so the radiation originated inside the uranium atom. This conclusion was harder to reach than it sounds, because some fraction of the radiation is absorbed in the material itself. This is why she used a thin layer, rather than a piece of the material under investigation.

The mineral pitchblende, rich in uranium oxide, was found to be more radioactive than pure uranium metal. Marie concluded that it must contain some other element that was more radioactive than uranium. In 1898, the Curies chemically separated out a new element they called polonium, after Marie's country of birth, Poland.

They measured its radioactivity as four hundred times more intense than that from pure uranium. Six months later, they chemically separated out from pitchblende another radioactive substance more than one *million* times more radioactive than uranium.

They called it radium. One gram of radium gives out enough energy to raise one gram of water from just above freezing to boiling in one hour. Where this energy could be coming from was a complete mystery. It was hard to imagine how the energy could be stores in the material-in a few days, radium gave off as much energy as the most potent chemical fuel, yet it kept on radiating year after year.

One popular theory at the time was that space was full of waves of energy, and almost all materials were completely transparent to this sea of radiation, but radium, polonium and uranium absorbed some of this energy and re-emitted it.

Bohr's Semiclassical Argument to Fix the Quantum of Angular Momentum

However, Bohr found a clever *theoretical* way to determine R_H from his model: by equating his prediction of the frequency emitted when an electron goes from one orbit to another in a very large atom with the classical prediction—which would be just the orbital frequency of the electron, how many times per second it goes around, he deduced $K = 1/2\pi$ and from that the Rydberg constant that appeared before is here given in terms of h, m and e.

The rather abstract argument that the quantum predictions must match the known classical results for large slow systems actually fixes the Rydberg constant.

His argument goes as follows: for the circular orbits,

$$\frac{mv^2}{r}=\frac{1}{4\pi\varepsilon_0}\cdot\frac{e^2}{r^2} \text{ so } mv^2=\frac{1}{4\pi\varepsilon_0}\cdot\frac{e^2}{r},\quad K.E.=-\tfrac{1}{2}P.E.,\quad E=-\frac{1}{4\pi\varepsilon_0}\cdot\frac{e^2}{2r}$$

With the angular momentum quantized, for the nth orbit:

$$mv_n r_n = nKh,$$

where h is Planck's constant, n an integer, K the unknown multiplying factor (ok, fixed by experiment, but we're finding it independently).

From this quantization condition we can find the radius, and hence the energy, of the nth orbit:

$$\frac{1}{4\pi\varepsilon_0}\cdot\frac{e^2}{r_n}=mv_n^2=m\left(\frac{nKh}{mr_n}\right)^2$$

Giving,

$$r_n=\frac{4\pi\varepsilon_0 n^2K^2h^2}{me^2},\quad E_n=-\frac{1}{4\pi\varepsilon_0}\cdot\frac{e^2}{2r_n}=-\left(\frac{1}{4\pi\varepsilon_0}\right)^2\cdot\frac{me^4}{2K^2h^2}\cdot\frac{1}{n^2}$$

In the large n limit,

$$E_{n+1}-E_n\cong\left(\frac{1}{4\pi\varepsilon_0}\right)^2\cdot\frac{me^4}{2K^2h^2}\cdot\frac{2}{n^3}=h\nu,$$

so,

$$\nu=\left(\frac{1}{4\pi\varepsilon_0}\right)^2\cdot\frac{me^4}{K^2h^3n^3}$$

where is the frequency of the emitted photon on jumping down one quantum number. In the classical limit of large n, ν *must* match the orbital frequency of the electron, since Maxwell's equations will be valid.

That is,

$$\nu=\frac{v_n}{2\pi r_n}=\frac{nKh}{2\pi mr_n^2}=\left(\frac{1}{4\pi\varepsilon_0}\right)^2\cdot\frac{nKh}{2\pi m}\cdot\frac{m^2e^4}{n^4k^4h^4}=\left(\frac{1}{4\pi\varepsilon_0}\right)^2\cdot\frac{me^4}{2\pi n^3K^3h^3}$$

Comparing the two expressions, we see that they agree if K = $1/2\pi$.

Putting K = $1/2\pi$ into the energy level formula,

$$E_n = -\left(\frac{1}{4\pi\varepsilon_0}\right)^2 \cdot \frac{me^4}{2K^2h^2} \cdot \frac{1}{n^2} = -\left(\frac{1}{4\pi\varepsilon_0}\right)^2 \cdot \frac{2\pi^2 me^4}{h^2} \cdot \frac{1}{n^2}$$

Now the Rydberg constant is defined by,

$$E_n = hcR_H/n^2$$

so the Bohr model *predicts* that

$$R_H = \left(\frac{1}{4\pi\varepsilon_0}\right)^2 \cdot \frac{2\pi^2 me^4}{ch^3}$$

This formula was found to be correct within the limits of experimental error in measuring the quantities on the right.But few people believed his theory. For one thing, it soon became apparent that in the spectra of some stars (actually including the sun) there were spectral lines apparently corresponding to *half* the angular momentum quantum. How could that be? Bohr's response was that these lines must be from ionized helium, not hydrogen.

A neutral helium atom has two electrons, a singly-ionized helium atom has just one electron, but the nucleus has a charge twice that of the hydrogen nucleus, so the factor e^4 in the Rydberg constant is replaced by $4e^4$, which leads to the observed result. But then a spectroscopist called Fowler did some very precise measurements, and found that actually the R_H for these new lines corresponded to a factor of 4.0016.

How could Bohr explain *that*? Bohr pointed out that at this level of precision, the finite mass of the nucleus must be taken into account by using a reduced mass for the electron. This gives just the right factor. This result greatly impressed Einstein, who concluded that Bohr must be on the right track.

The Bohr Atom is Still Important

Although, as we shall see shortly, Bohr's semiclassical analysis has long been replaced by Schrodinger's wave function, there are recent experiments in atomic physics where the classical approach provides valuable insight. In particular, so-called Rydberg atoms, which are atoms with one electron in a spatially large orbit (large n, weakly bound), act a lot like classical systems. Such atoms can be ionized by microwave fields.

For a considerable range of parameters, the onset of this ionization can be accounted for by ignoring quantum mechanics altogether, and interpreting ionization as the onset of chaotic motion in the *classical* driven system! (And, the standard perturbation theoretic methods of quantum mechanics don't work for this system anyway, because the perturbing microwave electric field is of the same order of magnitude as the atom's electric field at these large orbits.) We should mention that, counterintuitively, quantum mechanics *does* become important again at *very* large n (or high microwave frequency), where some tricks from condensed matter physics have been used successfully to interpret the experiments.

The next real advance in understanding the atom came from an unlikely quarter—a student prince in Paris. Prince Louis de Broglie was a member of an illustrious family, prominent in politics and the military since the 1600's. Louis began his university studies with history, but his elder brother Maurice studied x-rays in his own laboratory, and Louis became interested in physics. He worked with the very new radio telegraphy during the war.

After the war, de Broglie focused his attention on Einstein's two major achievements, the theory of special relativity and the quantization of light waves. He wondered if there could be some connection between them. Perhaps the quantum of radiation really should be thought of as a particle. It had been known for a long time that light waves carry momentum: this is famously demonstrated by the "radiometer", a small "windmill" in a vacuum, with vanes silver on one side and blackened on the other. If the vacuum is good, the radiometer begins to rotate when exposed to light because the light bouncing off the silvered side delivers twice the momentum of the light absorbed by the blackened side. (It should be added that cheap versions of this device have poor vacua, and the heated gas near the blackened side tends to push the vanes the wrong way.)

In fact, it follows from Maxwell's equations that the momentum density of a light beam is related to its energy density by $E = cp$. We would therefore expect this same energy-momentum relationship to be true for the photons of which the light beam is composed. Now, from special relativity we know that all particles have an energy-momentum relationship $E^2 = m_0^2c^4 + c^2p^2$, where m_0 is the rest mass of the particle. The only way this can be the same as $E = cp$ is if $m_0 = 0$, or, at least, if m_0 is so small that all our observations are on particles having kinetic energy so far in excess of their rest energy that the tiny mass is not detectable. De Broglie suspected that the photon *did* have a very tiny nonzero rest mass, so that if the speed of a sufficiently low energy quantum could be measured, it would be found to be less than c. On this point he was wrong (as far as we know!) Nevertheless, it was a very valuable conceptual breakthrough to think of the quantum of radiation as a *particle*, knowing full well that radiation is a wave. In fact, his incorrect idea that the photon (as we now call the light quantum) had a rest mass led him to analyse the relationship between particle properties and wave properties by transforming to the rest frame of the photon, and he discovered that the energy and momentum of the particle were related to the frequency and wavelength of the wave by:

$$B = hf,\ p = h/\lambda$$

Of course, the first condition is the Planck-Einstein quantization, and the second follows trivially from it if we take $E = cp$ and $\lambda f = c$ But de Broglie showed it was more generally true—it worked even if the photon had a rest mass.

Having decided that the photon might well be a particle with a rest mass, albeit very small, it dawned on de Broglie that in other respects it might not be too different from other particles, especially the very light electron. In particular, maybe the electron also had an associated wave. The obvious objection was that if the electron was wavelike, why had no diffraction or interference effects been observed? But there was an answer. If de Broglie's relation between momentum and wavelength $p = h/\lambda$, also held for electrons, the wavelength was sufficiently short that these effects would be easy to miss. As de Broglie himself pointed out, the wave nature of light isn't very evident in everyday life, or in ray tracing in geometrical optics. He suspected the apparently pure particle nature of electronic trajectories was analogous to the apparent straight-line propagation of rays of light, over distance scales much greater than the wavelength.

However, the wavelike properties should be important on an *atomic* scale. No progress had been made in a decade in understanding why the electronic orbits in the Bohr atom were restricted to integral values of the angular momentum in units of h. But if the electron were in some sense a wave, it would be very natural to restrict the orbits to those of standing waves, for otherwise the electron wave on going around the orbit would interfere with itself destructively.

Suppose now the electron, having momentum p, is moving in a circular orbit of radius r. Then for a standing wave, a whole number of wavelengths must fit around the circle, so for some integer n, $n\lambda = 2\pi r$ Putting this together with $p = h/\lambda$ we find:

$$2\pi r = n\lambda = nh/p$$

So,

$$L = pr = nh/2\pi$$

The "standing wave" condition immediately gives Bohr's quantization of angular momentum!

An Accident at the Phone Company Makes Everything Crystal Clear: There was an accident at the Bell Telephone Laboratories early nineties. Clinton Davisson and L. H. Germer, looking for ways to improve vacuum tubes, were watching how electrons from an electron gun in a vacuum tube scattered off a flat nickel surface. Suddenly, while the experiment was running and the nickel target was very hot, a bottle of liquid air near the apparatus exploded, smashing one of the vacuum pipes, and air rushed into the apparatus. The hot nickel target oxidized immediately. The layer of oxide made their target useless for further investigations. They decided to clean off the oxide by heating the nickel in a hydrogen atmosphere then in vacuum. After doing this for a prolonged period, the nickel looked good, and they resumed the investigation.

To their amazement, the pattern of electron scattering from the newly cleaned nickel target was completely different from that before the accident. What had changed? On examining their newly cleaned crystal carefully, they found a clue. The original target was polycrystalline—made up of a multitude of tiny crystals, oriented randomly. During the prolonged heating of the cleaning process, the nickel had re-crystallized into a few large crystals.

To quote from their paper: "It seemed probable to us from these results that the intensity of scattering from a single crystal would exhibit a marked dependence on crystal direction, and we set about at once preparing experiments for an investigation of this dependence. We must admit that the results obtained in these experiments have proved to be quite at variance with our expectations. It seemed likely that strong beams would be found issuing from the crystal along what may be termed its transparent directions—the directions in which the atoms in the lattice are arranged along the smallest number of lines per unit area. Strong beams are indeed found issuing from the crystal, but only when the speed of bombardment lies near one or another of a series of critical values, and then in directions quite unrelated to crystal transparency. "The most striking characteristic of these beams is a one to one orrespondence...which the strongest of them bear to the Laue beams that would be found issuing from the same crystal if the incident beam were a beam of x-rays. Certain others appear to be analogues... of optical diffraction beams from plane reflection gratings—the lines of these gratings being lines or rows of atoms in the surface of the crystal. Because of these similarities... a description... in terms of an equivalent wave radiation... is not only possible, but most simple and natural. This involves the association of a wavelength with the incident electron beam, and this wavelength turns out to be in acceptable agreement with the value h/mv of the undulatory mechanics, Planck's action constant divided by the momentum of the electron.

"That evidence for the wave nature of particle mechanics would be found in the reaction between a beam of electrons and a single crystal was predicted by Elsasser two years ago—shortly after the appearance of L. de Broglie's original papers on wave mechanics." It should be added that the two-slit diffraction pattern is of course exhibited by a beam of electrons, has been observed experimentally many times, and has precisely the same form as that for light. Electrons and photons generate interference patterns that are identical—although the short wavelength of the electrons used presents a challenge

Mechanical Waves

The definition of a wave: any disturbance of a system which displaces the system from its equilibrium state and has the characteristic that the disturbance can *propagate* from one part of the system to another. This seems like a quite general statement, and it is, but it can be reduced to a quite rigorous mathematical treatment which allows us to describe an incredibly large number of physical phenomena. Sound, light, ocean waves, radio and television transmission, earthquakes and more are all describable by the same mathematics which we refer to as wave mechanics. There are also inherent properties of all wave phenomena which can be observed. The most important of these is interference, the behaviour of a system when two or more waves occupy the same space at the same time.

For mechanical waves, we can be more restrictive in our definition and state that these disturbances we referred to must propagate in a medium, a material substance which is deformable and capable of transmitting a disturbance. Note that electromagnetic waves do *not* need a medium-they travel through empty space.

Quantitative Description of Mechanical Waves

Wave Function for a Sinusoidal Wave: The correspondence between uniform circular motion and transverse periodic wave motion leads to an immediate mathematical description of a so-called traveling wave, *i.e.* a periodic transverse wave with velocity v. For the wave as shown in figure, we have a wave function.

$$y(x,t) = A\sin\omega\ t + \frac{x}{v} + \phi = A\sin 2\pi f\ t + \frac{x}{v}\phi$$

for a wave moving in the -x direction. For a wave moving in the +x direction, change the +x/v sign to -x/*v* inside the parentheses. The angle ϕ is a phase angle and allows for a completely general description of the value of the displacement from equilibrium at $t = 0$. Note that this mathematical description yields the position *transverse* to the wave direction and relative to the equilibrium position of the medium. We get a more convenient form for the traveling sinusoidal wave if we make use of the wave number, *k*. Note that,

$$\omega = vk$$

So we can rewrite the expression for a wave moving in the +x direction as,

$$y(x,\ t) = A\ \sin(\omega t\text{-}kx + \phi)$$

We often refer to the phase of a wave. This is defined as,

$$\text{phase} = \omega t\ kx$$

The Wave Equation

We have defined a function which we can easily verify gives the position of a particle transverse to a traveling wave. The position is a function of wave position and time. We can find the first and second derivatives of this transverse position in order to get the particle velocity and particle acceleration transverse to the wave direction.

$$y(x,t) = A\sin(\omega t - kx + \phi)$$

$$v_y(x,t) = \frac{y(x,t)}{t} = \omega A\cos(\omega t - kx + \phi)$$

$$a_y(x,t) = \frac{y(x,t)}{t^2} = -\omega^2 A\sin(\omega t - kx + \phi)$$

where we have used partial derivatives with respect to time since the position is a function of position and time. Notice that for any given values of ϕ, x, and t, waiting a time interval of one period, $T = 2\pi/\omega$, yields the same values for the particle position, velocity, and acceleration. This is consistent with our definition of period. Notice also that the results for position, velocity, and acceleration are the same as those for a particle in simple harmonic motion with amplitude A and angular frequency ω. Now note something else. If we look at the second partial derivative of the particle position with respect to x, we have,

$$\frac{y(x, t)}{x^2} = -k^2 A \sin(\omega t - kx) = -k^2 y(x, t)$$

Using our derivation for the particle acceleration and the relation $\omega = v\ k$ gives,

$$\frac{y(x, t)/t^2}{y(x, t)/x^2} = \frac{-\omega^2 y(x, t)}{-k^2 y(x, t)} = \frac{\omega^2}{k^2} = v^2$$

Hence we derive the wave equation.

$$\frac{y(x, t)}{x^2} = \frac{1}{v^2} \frac{y(x, t)}{t^2}$$

Energy and Power of a Traveling Wave on a String

The Rate of Energy Transmission

As the wave moves along the x direction, forces due to tension in the string continuously do work to transfer kinetic and potential energy from one part of the string to another. The kinetic energy dK associated with a string element dm is given by,

$$dk = \frac{1}{2} dm\, u^2$$

where u is the transverse speed of the oscillating string element. We use the velocity from equation to find u.

$$u = \frac{y(x,t)}{t} = \omega A \cos(\omega t - kx + \phi)$$

In terms of the mass density, $dm = \mu\ dx$, we have,

$$dK = \frac{1}{2}(\mu dx)(\omega A)^2 \cos^2(wt - kx + \phi)$$

Divide this quantity by dt to get the rate at which kinetic energy is transported as,

$$\frac{dK}{dt} = \frac{1}{2}\mu\frac{dx}{dt}\omega^2 A^2 \cos^2(\omega t - kx + \phi)$$
$$= \mu v\omega^2 A^2 \cos^2(\omega t - kx + \phi).$$

This is the instantaneous power transferred by the transverse wave. The *average* rate of kinetic energy transfer is,

$$\frac{dK}{dt_{avg}} = \frac{1}{2}\mu v\omega^2 A^2\left[\cos^2(\omega t - kx + \phi)\right]_{avg.} = \frac{1}{4}\mu v\omega^2 A^2$$

for simple harmonic systems, the average total energy is just twice the average kinetic energy (*i.e.* the average potential energy equals the average kinetic energy). Therefore, the average power transmitted by the wave is,

$$P_{avg.} = 2\frac{dK}{dt_{avg}} = \frac{1}{2}\mu v\omega^2 A^2$$

The average power depends on the square of the amplitude and the square of the angular frequency.

This result is general and true for waves of all types. By similar arguments, we can show that a traveling wave also carries linear momentum. We shall explicitly derive this for the case of electromagnetic waves.

Superposition of Mechanical Waves

Two or more waves can pass simultaneously through the same region of space. We can ask what the result of this is. Waves follow the principle of superposition which says that the net effect of the waves is the sum of their individual effects.

Overlapping waves do not alter the travel of each other. The mathematics for describing overlapping waves of the same frequency and wavelength, we call this interference, is not difficult. For example, imagine two transverse waves with the same amplitude, frequency, direction, and wavelength traveling along a string.

We can specify different spatial positions for these waves through different phase angles. For simplicity, we assume the phase angle of one of these waves is zero as this does not change the mathematics or physics of the description. According to the principle of superposition, the overlap of these two waves can be described as follows.

$$\begin{aligned} y_{net}(x,t) &= y_1(x,t) + y_2(x,t) \\ &= A\sin(\omega t - kx) + A\sin(\omega - kxt + \phi) \\ y_{net} &= \left[2A\cos\frac{1}{2}\phi\right]\sin\left(\omega - kx + \frac{1}{2}\phi\right). \end{aligned}$$

If we plot this equation, we see that it is also a sinusoidal wave traveling in the +x direction. This net wave is the *only* wave you would see traveling along the string. The two interfering waves would not be visible. We note that by changing the value of f we can get a resultant wave which shows constructive interference or destructive interference. For completely constructive interference, we want crests of one wave to coincide with crests of the other and troughs to coincide with troughs. For completely destructive interference, we want crests of one wave to coincide with troughs of the other, thereby leading to complete cancellation of the net wave. We see examples of these cases in figure.

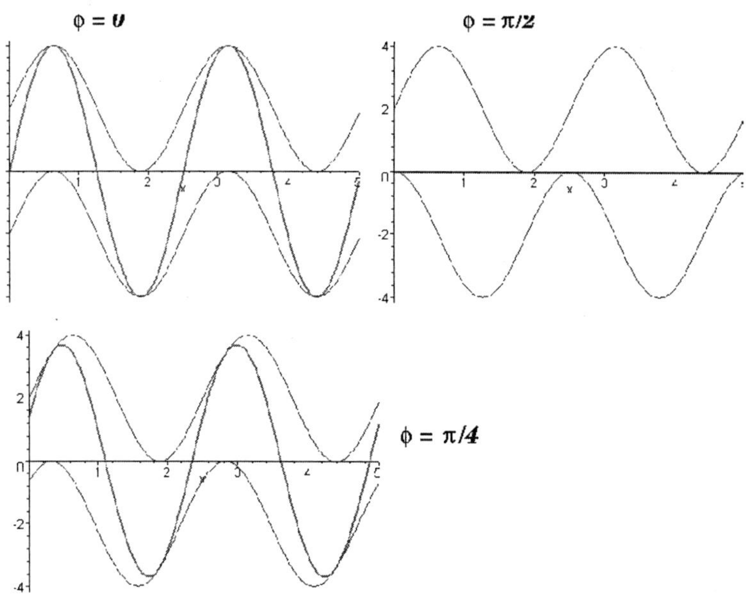

Figure: Two Waves with the Same Amplitude and Frequency but Independent Phase Angles are Depicted in Blue.

The top curve has phase angle equal to zero and the bottom curve has the phase angle set as shown by the label of the graph.The red curve is the result of superposing the two waves.

Chapter 4

Quantum Numbers

Quantum numbers describe values of conserved numbers in the dynamics of the quantum system. They often describe specifically the energies of electrons in atoms, but other possibilities include angular momentum, spin etc. Since any quantum system can have one or more quantum numbers, it is a futile job to list all possible quantum numbers. The question of *how many quantum numbers are needed to describe any given system* has no universal answer, although for each system one must find the answer for a full analysis of the system. The dynamics of any quantum system are described by a quantum Hamiltonian, H.

There is one quantum number of the system corresponding to the energy, *i.e.*, the eigenvalue of the Hamiltonian. There is also one quantum number for each operator O that commutes with the Hamiltonian (*i.e.* satisfies the relation OH = HO). These are all the quantum numbers that the system can have. Note that the operators O defining the quantum numbers should be independent of each other. Often there is more than one way to choose a set of independent operators. Consequently, in different situations different sets of quantum numbers may be used for the description of the same system.

Single Electron in an Atom

The most widely studied set of quantum numbers is that for a single electron in an atom: because it is not only useful in chemistry, being the basic notion behind the periodic table, valence and a host of other properties, but also because it is a solvable and realistic problem, and, as such, finds widespread. In non-relativistic quantum mechanics the Hamiltonian of this system consists of the kinetic energy of the electron and the potential energy due to the Coulomb force between the nucleus and the electron.

The kinetic energy can be separated into a piece which is due to angular momentum, J, of the electron around the nucleus, and the remainder. Since the potential is spherically symmetric, the full Hamiltonian commutes with J^2. J^2 itself commutes with any one of the components of the angular momentum vector, conventionally taken to be J_z. These are the only mutually commuting operators in this problem; hence, there are three quantum numbers.

These are conventionally known as:

- The principal quantum number (n = 1, 2, 3, 4...) denotes the eigenvalue of H with the J^2 part removed. This number therefore has a dependence only on the distance between the electron and the nucleus (ie, the radial coordinate, r). The average distance increases with n, and hence quantum states with different principal quantum numbers are said to belong to different shells.
- The azimuthal quantum number (l = 0, 1... n"1) (also known as the angular quantum number or orbital quantum number) gives the orbital angular momentum through the relation $L^2 = h^2 l(l + 1)$. In chemistry, this quantum number is very important, since it specifies the shape of an atomic orbital and strongly influences chemical bonds and bond angles. In some contexts, l=0 is called an s orbital, l=1, a p orbital, l=2, a d orbital and l=3, an f orbital.
- The magnetic quantum number (m_l = "l, "l+1... 0... l"1, l) is the eigenvalue, $L_z = m_l$h. This is the projection of the orbital angular momentum along a specified axis.

Results from spectroscopy indicated that up to two electrons can occupy a single orbital. However two electrons can never have the same exact quantum state nor the same set of quantum numbers according to Hund's Rules, which addresses the Pauli exclusion principle.

A fourth quantum number with two possible values was added as an *ad hoc* assumption to resolve the conflict; this supposition could later be explained in detail by relativistic quantum mechanics and from the results of the renown Stern-Gerlach experiment.

- The spin projection quantum number (m_s = "1/2 or +1/2), the intrinsic angular momentum of the electron. This is the projection of the spin s=1/2 along the specified axis.

Example: The quantum numbers used to refer to the outermost valence electron of the Fluorine (F) atom, which is located in the 2p atomic orbital, are; $n = 2$, $l = 1$, $m_l = 1$, or 0, or –1, $m_s = -1/2$ or 1/2. Note that molecular orbitals require totally different quantum numbers, because the Hamiltonian and its symmetries are quite different.

Quantum Numbers with Spin-orbit Interaction

When one takes the spin-orbit interaction into consideration, l, m and s no longer commute with the Hamiltonian, and their value therefore changes over time. Thus another set of quantum numbers should be used.

This set includes:

- The total angular momentum quantum number (j = 1/2,3/2... n"1/2) gives the total angular momentum through the relation $J^2 = h^2 j(j + 1)$.
- The projection of the total angular momentum along a specified axis (m_j = -j,-j+1... j), which is analogous to m, and satisfies $m_j = m_l + m_s$.
- Parity. This is the eigenvalue under reflection, and is positive (*i.e.* +1) for states which came from even l and negative (*i.e.* -1) for states which came from odd l. The former is also known as even parity and the latter as odd parity

For example, consider the following eight states, defined by their quantum numbers:

- $l = 1$, $m_l = 1$, $m_s = +1/2$
- $l = 1$, $m_l = 1$, $m_s = -1/2$
- $l = 1$, $m_l = 0$, $m_s = +1/2$
- $l = 1$, $m_l = 0$, $m_s = -1/2$
- $l = 1$, $m_l = -1$, $m_s = +1/2$
- $l = 1$, $m_l = -1$, $m_s = -1/2$
- $l = 0$, $m_l = 0$, $m_s = +1/2$
- $l = 0$, $m_l = 0$, $m_s = -1/2$

The quantum states in the system can be described as linear combination of these eight states. However, in the presence of spin-orbit interaction, if one wants to describe the same system by eight states which are eigenvectors of the Hamiltonian, we should consider the following eight states:

- $j = 3/2$, $m_j = 3/2$, odd parity (coming from state (1))
- $j = 3/2$, $m_j = 1/2$, odd parity (coming from states (2) and (3))
- $j = 3/2$, $m_j = -1/2$, odd parity (coming from states (4) and (5))
- $j = 3/2$, $m_j = -3/2$, odd parity (coming from state (6))
- $j = 1/2$, $m_j = 1/2$, ódd parity (coming from state (2) and (3))
- $j = 1/2$, $m_j = -1/2$, odd parity (coming from states (4r) and (5))
- $j = 1/2$, $m_j = 1/2$, even parity (coming from state (7))
- $j = 1/2$, $m_j = -1/2$, even parity (coming from state (8))

Elementary Particles

For a more complete description of the quantum states of elementary particles. Elementary particles contain many quantum numbers which are usually said to be intrinsic to them. However, it should be understood that the elementary particles are quantum states of the standard model of particle physics, and hence the quantum numbers of these particles bear the same relation to the Hamiltonian of this model as the quantum numbers of the Bohr atom does to its Hamiltonian.

In other words, each quantum number denotes a symmetry of the problem. It is more useful in field theory to distinguish between spacetime and internal symmetries. Typical quantum numbers related to spacetime symmetries are spin the parity, C-parity and T-parity. Typical internal symmetries are lepton number and baryon number or the electric charge. It is worth mentioning here a minor but often confusing point. Most conserved quantum numbers are additive. Thus, in an elementary particle reaction, the sum of the quantum numbers should be the same before and after the reaction. However, some, usually called a *parity*, are multiplicative; ie, their product is conserved. All multiplicative quantum numbers belong to a symmetry in which applying the symmetry transformation twice is equivalent to doing nothing. These are all examples of an abstract group called Z_2.

Quantum Mechanics

Quantum mechanics is a physical science dealing with the behaviour of matter and energy on the scale of atoms and subatomic particles/waves. Quantum mechanics is the foundation of several related disciplines, including condensed matter physics, quantum chemistry, and particle physics, and is the chief investigative tool in structural biology. Within the field of engineering, quantum mechanics plays an important part of forming the development of nanotechnology and electronics.The term "quantum mechanics" was first coined by Max Born in 1924.

The acceptance by the general physics community of quantum mechanics is due to its accurate prediction of the physical behaviour of systems, including systems where Newtonian mechanics fails. Even general relativity is limited—in ways quantum mechanics is not—in describing systems at the atomic scale or smaller, at very low or very high energies, or at the lowest temperatures. Through a century of experimentation and applied science, quantum mechanical theory has proven to be very successful and practical.

The Unexpected

Towards the end of the nineteenth century it seemed that physics had been very nearly perfected, but then paradox after paradox radically changed human concepts at fundamental levels. The models by which humans explained things on their scale in space and time failed to apply to the ultra small, the ultra massive, and to interactions at the highest speeds. A view of the universe that had seemed to be the distillation of commonplace observations was challenged by observations and theories that predicted correctly where classical physics had given impossible results. But the emerging picture was of a universe that stubbornly refused to behave the way humans had come to expect.

On the large end of the scale, relativity theory informed humans that time does not pass at the same rate for all observers, that matter can convert to energy and vice-versa, that two objects each moving at more than half the speed of light cannot have a closing speed that is equal to or greater than the speed of light, that time progresses more slowly near massive objects, etc. Things just do not work the way that experience with scales, rulers, and clocks here on earth had led humans to expect.

On the small end of the scale, wonders are even more abundant. A thing such as a photon or an electron does not have location and a traceable trajectory between the point where it is emitted and the point where it is detected. The points where such a particle may be detected are not what one would expect based on everyday experience. With a certain small probability, the detection point can even be on the other side of a solid barrier.

Probability turns out to be a highly salient factor in all interactions on this scale. The trajectory of any atomic scale object is "squishy" in the sense that any measurement intended to get clearer on the object's position forces knowledge of its velocity to become less clear, and vice-versa.In the era of classical physics, Newton and his followers believed that light consisted of particles, and others believed that light was actually a series of wave fronts propagated through some medium.

Rather than finding an experiment that would prove one or the other side correct, physicists learned that making an experiment designed to show light's frequency or other wave *characteristics* will manifest its "wave nature," and making an experiment that is designed to show its momentum or other particle characteristics will manifest its "particle nature." Furthermore, objects as large as atoms, and even some molecules, have shown their "wave nature" when addressed in the appropriate way.The most eminent of physicists have warned that if one imagines some explanation of quantum physics to "make sense," then that explanation is very likely to be flawed.

How the Unexpected Came to Light

The foundations of quantum mechanics date from the early 1800s, but the real beginnings of QM date from the work of Max Planck in 1900. Albert Einstein and Niels Bohr soon made important contributions to what is now called the "old quantum theory." However, it was not until 1924 that a more complete picture emerged with Louis de Broglie's matter-wave hypothesis and the true importance of quantum mechanics became clear. Some of the most prominent scientists to subsequently contribute in the mid-1920s to what is now called the "new quantum mechanics" or "new physics" were Max Born, Paul Dirac, Werner Heisenberg, Wolfgang Pauli, and Erwin Schrödinger.

Later, the field was further expanded with work by Julian Schwinger, Sin-Itiro Tomonaga and Richard Feynman for the development of Quantum Electrodynamics in 1947 and by Murray Gell-Mann in particular for the development of Quantum chromodynamics.

The interference that produces coloured bands on bubbles cannot be explained by a model that depicts light as a particle. It can be explained by a model that depicts it as a wave. The drawing shows sine waves that resemble waves on the surface of water being reflected from two surfaces of a film of varying width, but that depiction of the wave nature of light is only a crude analogy.Early researchers differed in their explanations of the fundamental nature of what we now call electromagnetic radiation. Some maintained that light and other frequencies of electromagnetic radiation are composed of particles, while others asserted that electromagnetic radiation is a wave phenomenon.

In classical physics these ideas are mutually contradictory. Ever since the early days of QM, scientists have acknowledged that neither idea by itself can explain electromagnetic radiation.In 1690, Christiaan Huygens explained the laws of reflection and refraction on the basis of a wave theory. Sir Isaac Newton believed that light consisted of infinitesimally small particles which he designated "corpuscles". In 1827 Thomas Young and Augustin Fresnel made experiments on interference that showed that a corpuscular theory of light was inadequate. Then in 1873 James Clerk Maxwell showed that by making an electrical circuit oscillate it should be possible to produce electromagnetic waves. His theory made it possible to compute the speed of electromagnetic radiation purely on the basis of electrical and magnetic measurements, and the computed value corresponded very closely to the empirically measured speed of light. In 1888, Heinrich Hertz made an electrical device that actually produced what we would now call microwaves — essentially radiation at a lower frequency than visible light.

Everything up to that point suggested that Newton had been entirely wrong to regard light as corpuscular.Later experiments indicated that a packet or quantum model was needed to explain some phenomena. When light strikes an electrical conductor it causes electrons to move away from their original positions. The observed phenomenon could only be explained by assuming that the light delivers energy in definite packets. In a photoelectric device such as the light meter in a camera, light hitting the metallic detector causes electrons to move. Greater intensities of light at one frequency can cause more electrons to move, but they will not move faster. In contrast, higher frequencies of light can cause electrons to move faster.

Therefore, intensity of light controls current, but frequency of light controls voltage. These observations raised a contradiction when compared with sound waves and ocean waves, where only intensity was needed to predict the energy of the wave. In the case of light, frequency appeared to predict energy. Something was needed to explain this phenomenon and to reconcile experiments that had shown light to have particle nature with experiments that had shown it to have wave nature.

Spectroscopy and Onward

It is fairly easy to see a spectrum produced by white light when it passes through a prism, the bevelled edge of a mirror or a tapered pane of glass, or through drops of rain to form a rainbow.

When samples which are composed purely of one single chemical element, *e.g.* hydrogen, are caused to emit light, they may emit light at several characteristic frequencies. The frequency profile produced is characteristic of that element. Instead of there being a wide band filled with colours from violet to red, there will be isolated bands of single colours separated by darkness.

Such a display is called a line spectrum. Some lines go beyond the visible frequencies and can only be detected by special photographic film or other such devices.

Scientists hypothesized that an atom could radiate light the way the string on a fine violin radiates sound – not only with a fundamental frequency but with several higher harmonics (formed when the string divides itself into halves and other divisions that vibrate in coordination with each other as when one half of the string is going one way as the other half of the string is going the opposite way). For a long time nobody could find a mathematical way to relate the frequencies of the line spectrum of any element.

In 1885, Johann Jakob Balmer (1825-1898) figured out how the frequencies of atomic hydrogen are related to each other. Where λ is wavelength, R is the Rydberg constant and n is an integer (n =3, 4,...) This formula can be generalized to apply to atoms that are more complicated than hydrogen, but we will stay with hydrogen for this general exposition. The next development was the discovery of the Zeeman effect, named after Pieter Zeeman (1865-1943). The physical explanation of the Zeeman effect was worked out by Hendrik Antoon Lorentz (1853-1928). Lorentz hypothesized that the light emitted by hydrogen was produced by vibrating electrons.

It was possible to get feedback on what goes on within the atom because moving electrons create a magnetic field and so can be influenced by the imposition of an external magnetic field in a manner analogous to the way that one iron magnet will attract or repel another magnet.

The Zeeman effect could be interpreted to mean that light waves are originated by electrons vibrating in their orbits, but classical physics could not explain why electrons should not fall out of their orbits and into the nucleus of their atoms, nor could classical physics explain why their orbits would be such as to produce the series of discrete frequencies derived by Balmer's formula and displayed in the line spectra. Why did the electrons not produce a continuous spectrum?

Old Quantum Theory

Quantum mechanics developed from the study of electromagnetic waves through spectroscopy which includes visible light seen in the colours of the rainbow, but also other waves including the more energetic waves like ultraviolet light, x-rays, and gamma rays; and the waves with longer wavelengths including infrared waves, microwaves and radio waves.

Only waves that travel at the speed of light are included in this description. Also, when the word "particle" is used below, it always refers to elementary or subatomic particles.

Planck's Constant

Classical physics predicted that a black-body radiator would produce infinite energy, but that result was not observed in the laboratory. If black-body radiation was dispersed into a spectrum, then the amount of energy radiated at various frequencies rose from zero at one end, peaked at a frequency related to the temperature of the radiating object, and then fell back to zero. In 1900, Max Planck developed an empirical equation that could account for the observed energy curves, but he could not harmonize it with classical theory. He concluded that the classical laws of physics do not apply on the atomic scale as had earlier been assumed. In this theoretical account, Planck allowed all possible frequencies, all possible wavelengths. However, he restricted the energy that is delivered. "In classical physics,... the energy of a given oscillator depends merely on its amplitude, and this amplitude is subject to no restriction." But, according to Planck's theory, the energy emitted by an oscillator is strictly proportional to its frequency.

The higher the frequency, the greater the energy. To reach this theoretical conclusion, he postulated that a radiating body consisted of an enormous number of elementary oscillators, some vibrating at one frequency and some at another, with all frequencies from zero to infinity being represented. The energy E of any one oscillator was not permitted to take on any arbitrary value, but was proportional to some integral multiple of the frequency f of the oscillator.

That is,

$$E = nhf.$$

where n =1, 2, 3,... The proportionality constant h is called *Planck's constant.*

One of the most direct applications is finding the energy of photons. If h is known, and the frequency of the photon is known, then the energy of the photons can be calculated. For instance, if a beam of light illuminated a target, and its frequency was 540×10^{12} hertz, then the energy of each photon would be $h \times 540 \times 10^{12}$ joules. The value of h itself is exceedingly small, about $6.6260693 \times 10^{-34}$ joule seconds. This means that the photons in the beam of light have an energy of about 3.58×10^{-19} joules or approximately 2.23 eV.

When the energy of a wave is described in this manner, it seems that the wave is carrying its energy in little packets. This discovery then seemed to remake the wave into a particle. These packets of energy carried along with the wave were called quanta by Planck. Quantum mechanics began with the discovery that energy is delivered in packets whose size is related to the frequencies of all electromagnetic waves (and to the colour of visible light since in that case frequency determines colour). Be aware, however, that these descriptions in terms of packet, wave and particle import macro-world concepts into the quantum world, where they have only provisional relevance or appropriateness.

In early research on light, there were two competing ways to describe light, either as a wave propagated through empty space, or as small particles traveling in straight lines. Because Planck showed that the energy of the wave is made up of packets, the particle analogy became favoured to help understand how light delivers energy in multiples of certain set values designated as quanta of energy. Nevertheless, the wave analogy is also indispensable for helping to understand other light phenomena.

In 1905, Albert Einstein used Planck's constant to explain the photoelectric effect by postulating that the energy in a beam of light occurs in concentrations that he called light quanta, later on called photons. According to that account, a single photon of a given frequency delivers an invariant amount of energy. In other words, individual photons can deliver more or less energy, but only depending on their frequencies. Although the description that stemmed from Planck's research sounds like Newton's corpuscular account, Einstein's photon was still said to have a frequency, and the energy of the photon was accounted proportional to that frequency. The particle account had been compromised once again.

Both the idea of a wave and the idea of a particle are models derived from our everyday experience. We can only investigate their properties indirectly. We look at some phenomena, such as the rainbow of colours that we see when a thin film of oil rests on the surface of a puddle of water, and we can explain that phenomenon to ourselves by comparing light with waves. We look at other phenomena, such as the way a photoelectric meter in our camera works, and we explain it by analogy to particles colliding with the detection screen in the meter. In both cases we take concepts from our everyday experience and apply them to a world we have never seen.

Neither form of explanation, wave or particle, is entirely satisfactory. In general any model can only approximate that which it models. A model is useful only within the range of conditions where it is able to predict the real thing with accuracy. Newtonian physics is still a good predictor of many of the phenomena in our everyday life. To remind us that both "wave" and "particle" are concepts imported from our macro world to explain the world of atomic-scale phenomena, some physicists such as Banesh Hoffmann have used the term "wavicle" to refer to whatever it is that is really there.

Reduced Planck's Constant

Planck's constant originally represented the energy that a light wave carries as a function of its frequency. A step in the development of this concept appeared in Bohr's work. Bohr was using a "planetary" or particle model of the electron, and could not understand why a 2π factor was essential to his experimentally derived formulae.

Later, de Broglie postulated that electrons have frequencies, just as do photons, and that the frequency of an electron must conform to the conditions for a standing wave that can exist in a certain orbit.

That is to say, the beginning of one cycle of a wave at some point on the circumference of a circle must coincide with the end of some cycle. There can be no gap, no length along the circumference that is not participating in the vibration, and there can be no overlap of cycles. So the circumference of the orbit, C, must equal the wavelength,

λ, of the electron multiplied by some positive integer ($n = 1, 2, 3...$). Knowing the circumference one can calculate wavelengths that fit that orbit, and knowing the radius, r, of the orbit one can calculate its circumference.

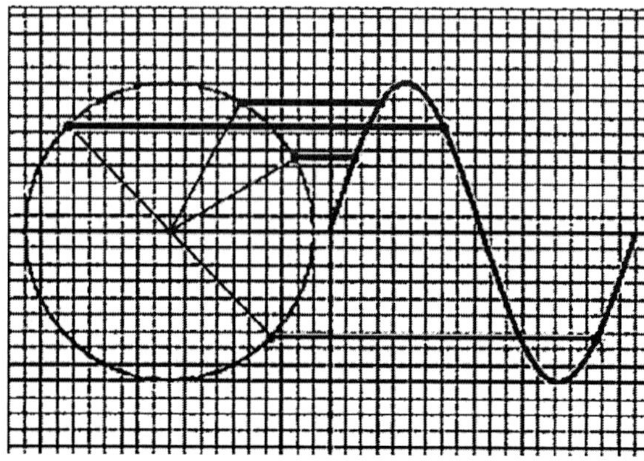

Figure: Relation between a Cycle and a Wave

To put all that in mathematical form,

$$C = 2\pi r = n\lambda$$

and so,

$$\lambda = 2\pi r/n$$

and the appearance of the 2π factor is seen to occur simply because it is needed to calculate possible wavelengths (and therefore possible frequencies) when the radius of an orbit is already known.

Again in 1925 when Werner Heisenberg developed his full quantum theory, calculations involving wave analysis called Fourier series were fundamental, and so the "reduced" version of Planck's constant ($h/2\pi$) became invaluable because it includes a conversion factor to facilitate calculations involving wave analysis. Finally, when this reduced Planck's constant appeared naturally in Dirac's equation it was then given an alternate designation, "Dirac's constant." Therefore, it is appropriate to begin with an explanation of what this constant is, even though the theories that made its use convenient.

As noted above, the energy of any wave is given by its frequency multiplied by Planck's constant. A wave is made up of crests and troughs. In a wave, a cycle is defined by the return from a certain position to the same position such as from the top of one crest to the next crest. A cycle actually is mathematically related to a circle, and both have 360 degrees. A degree is a unit of measure for the amount of turn needed to produce an arc of a certain length at a given distance. A sine curve is generated by a point on the circumference of a circle as that circle rotates. There are 2π radians per cycle in a wave, which is mathematically related to the way a circle has 360° (which are equal to two π radians). (A radian is simply the angle if a distance along the circumference of the circle is measured equal to the radius of the circle, and then lines are drawn to the centre of the circle. This forms an angle equal to 1 radian.) Since one cycle is 2π radians, when h is divided by 2π the two "2 π" factors will cancel out leaving just the radian to contend with. So, dividing h by 2π describes a constant that, when multiplied by the frequency of a wave, gives the energy in joules per radian. The reduced Planck's constant is written in mathematical formula7e as ‘, and is read as "h-bar".

$$\hbar = \frac{h}{2\pi}.$$

The reduced Planck's constant allows computation of the energy of a wave in units per radian instead of in units per cycle. These two constants h and ‘are merely conversion factors between energy units and frequency units.

The reduced Planck's constant is used more often than h (Planck's constant) alone in quantum mechanical mathematical formulas for many reasons, one of which is that angular velocity or angular frequency is ordinarily measured in radians per second so using ‘ that works in radians too will save a computation to put radians into degrees or vice-versa. Also, when equations relevant to those problems are written in terms of ‘, the frequently occurring 2π factors in numerator and denominator can cancel out, saving a computation. However, in other cases, as in the orbits of the Bohr atom, $h/2\pi$ was obtained naturally for the angular momentum of the orbits. Another expression for the relation between energy and wave length is given in electron volts for energy and angstroms for wavelength:

$$E_{\text{photon}}\ (\text{eV}) = 12{,}400/\lambda(\text{Å})$$

it appears not to involve h at all, but that is only because a different system of units has been used and now, numerically, the appropriate conversion factor is 12,400.

Bohr Atom

In 1897 the particle called the electron was discovered. By means of the gold foil experiment physicists discovered that matter is, volume for volume, largely space. Once that was clear, it was hypothesized that negative charge entities called electrons surround positively charged nuclei. So at first all scientists believed that the atom must be like a miniature solar system. But that simple analogy predicted that electrons would, within about one hundredth of a microsecond, crash into the nucleus of the atom.

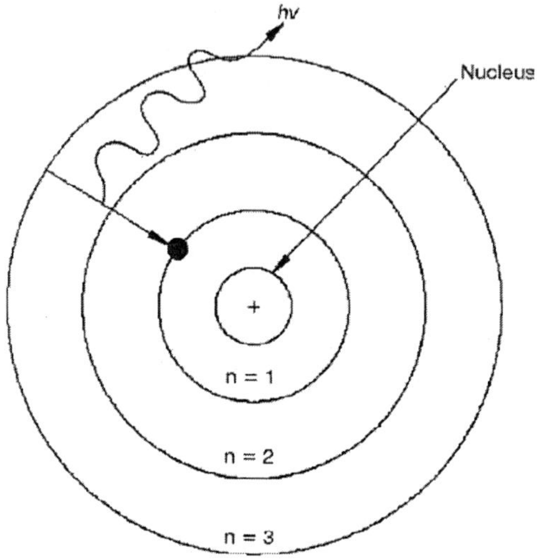

Figure: Bohr Model of the Atom

The great question of the early 20th century was, "Why do electrons normally maintain a stable orbit around the nucleus?"

In 1913, Niels Bohr removed this substantial problem by applying the idea of discrete (non-continuous) quanta to the orbits of electrons. This account became known as the Bohr model of the atom. Bohr basically theorized that electrons can only inhabit certain orbits around the atom. These orbits could be derived by looking at the spectral lines produced by atoms.

Bohr explained the orbits that electrons can take by relating the angular momentum of electrons in each "permitted" orbit to the value of h, Planck's constant. He held that an electron in the lowest orbital has a discrete angular momentum equal to $h/2\pi$.

Each orbit after the initial orbit must provide for an electron's angular momentum being an integer multiple of that lowest value. He depicted electrons in atoms as being analogous to planets in a solar orbit. However, he took Planck's constant to be a fundamental quantity that introduces special requirements at this subatomic level and that explains the spacing of those "planetary" orbits.

Bohr considered one revolution in orbit to be equivalent to one cycle in an oscillator (as in Planck's initial measurements to define the constant h) which is in turn similar to one cycle in a wave.

The number of revolutions per second is (or defines) what we call the frequency of that electron or that orbital. Specifying that the frequency of each orbit must be an integer multiple of Planck's constant h would only permit certain orbits, and would also fix their size.

Bohr generalized Balmer's formula for hydrogen by replacing denominator in the term 1/4 with an explicit squared variable:

$$\frac{1}{\lambda} = R_H\left(\frac{1}{m^2} - \frac{1}{n^2}\right), \quad m = 1, 2, 3, 4, 5, \ldots, \text{ and } n > m$$

where λ is the wavelength of the light, R_H is the Rydberg constant for hydrogen, and the integers n and m refer to the orbits between which electrons can transit. This generalization predicted many more line spectra than had been previously detected, and experimental confirmation of this prediction followed. It follows almost immediately that if λ is quantized as the formula above indicates, then the momentum of any photon must be quantized.

The frequency of light, ν, at a given wavelengtλ is given by the relationship:

$$\upsilon = \frac{c}{\lambda} \text{ and}$$

$$\lambda = \frac{c}{\nu} \text{ and}$$

multiplying by h/h = 1,

$$\lambda - \frac{hc}{h\nu}, \text{ and we know that } E = hí \text{ so}$$

which we can rewrite as:

$$\lambda = \frac{hc}{E}$$

$$\lambda = \frac{h}{E/c}, \text{ and } E/c = p \text{ (momentum) so}$$

$$\lambda = \frac{h}{p} \text{ or}$$

$$p = \frac{h}{\lambda}.$$

Beginning with line spectra, physicists were able to deduce empirically the rules according to which the orbits of electrons are determined and to discover something vital about the momentums involved that they are quantized.

Bohr next realised how the angular momentum of an electron in its orbit, L, is quantized, *i.e.*, he determined that there is some constant value K such that when it is multiplied by Planck's constant, h, it will yield the angular momentum that pertains to the lowest orbital. When it is multiplied by successive integers it will then give the values of other possible orbitals. He later determined that $K = 1/2\pi$.

Bohr's theory represented electrons as orbiting the nucleus of an atom in a way that was amazingly different from what we see in the world of our everyday experience. He showed that when an electron changed orbits it did not move in a continuous trajectory from one orbit around the nucleus to another. Instead, it suddenly disappeared from its original orbit and reappeared in another orbit. Each distance at which an electron can orbit is a function of a quantized amount of energy.

The closer to the nucleus an electron orbits, the less energy it takes to remain in that orbital. Electrons that absorb a photon gain a quantum of energy, so they jump to an orbit that is farther from the nucleus, while electrons that emit a photon lose a quantum of energy and so jump to an inner orbital. Electrons cannot gain or lose a fractional quantum of energy, and so, it is argued, they cannot have a position that is at a fractional distance between allowed orbitals.

Allowed orbitals were designated as whole numbers using the letter n with the innermost orbital being designated $n = 1$, the next out being $n = 2$, and so on. Any orbital with the same value of n is called an electron shell.

Bohr's model of the atom was essentially two-dimensional because it depicts electrons as particles in circular orbits. In this context, two-dimensional means something that can be described on the surface of a plane. One-dimensional means something that can be described by a line. Because circles can be described by their radius, which is a line segment, sometimes Bohr's model of the atom is described as one-dimensional.

Wave-particle Duality

Niels Bohr determined that it is impossible to describe light adequately by the sole use of either the wave analogy or of the particle analogy. Therefore he enunciated the principle of complementarity, which is a theory of pairs, such as the pairing of wave and particle or the pairing of position and momentum. Louis de Broglie worked out the mathematical consequences of these findings. In quantum mechanics, it was found that electromagnetic waves could react in certain experiments as though they were particles and in other experiments as though they were waves. It was also discovered that subatomic particles could sometimes be described as particles and sometimes as waves. This discovery led to the theory of wave-particle duality by Louis-Victor de Broglie in 1924, which states that subatomic entities have properties of both waves and particles at the same time.

The Bohr atom model was enlarged upon with the discovery by de Broglie that the electron has wave-like properties. In accord with de Broglie's conclusions, electrons can only appear under conditions that permit a standing wave. A standing wave can be made if a string is fixed on both ends and made to vibrate. That illustration shows that the only standing waves that can occur are those with zero amplitude at the two fixed ends.

The waves created by a stringed instrument appear to oscillate in place, simply changing crest for trough in an up-and-down motion. A standing wave can only be formed when the wave's length fits the available vibrating entity. In other words, no partial fragments of wave crests or troughs are allowed. In a round vibrating medium, the wave must be a continuous formation of crests and troughs all around the circle. Each electron must be its own standing wave in its own discrete orbital.

Development of Modern Quantum Mechanics

Full Quantum Mechanical Theory

Werner Heisenberg made the full quantum mechanical theory in 1925 at the young age of 23. Following his mentor, Niels Bohr, Werner Heisenberg began to work out a theory for the quantum behaviour of electron orbitals. Because electrons could not be observed in their orbits, Heisenberg went about creating a mathematical description of quantum mechanics built on what could be observed, that is, the light emitted from atoms in their characteristic atomic spectra. Heisenberg studied the electron orbital on the model of a charged ball on a spring, an oscillator, whose motion is anharmonic. For a picture of the behaviour of a charged ball on a spring see: Vibrating Charges. Heisenberg first explained this kind of observed motion in terms of the laws of classical mechanics known to apply in the macro world, and then applied quantum restrictions, discrete (non-continuous) properties, to the picture.

Doing so causes gaps to appear between the predicted orbitals so that the mathematical description he formulated would then represent only the electron orbitals predicted on the basis of the atomic spectra.

In approaching the problem that Bohr gave him to solve, Heisenberg took the strategic stance that he would not deal with unobservable quantities. He would begin formulating equations using only the quantities that could be observed. That strategy led him to begin with the actual experimental evidence at hand: Measurements had been well established for such data as:

- The frequencies (and the mathematically related energies) emitted or absorbed by electron transitions from one of the Bohr stationary orbits, known to be associated with the bright line spectra.
- The "transition amplitude" or likelihood of transition from any given orbit to any given orbit, known from the strength of the various lines in the bright spectrum, etc.

From classical formulas that would characterize those phenomena Heisenberg created analogous formulas that took account of quantum conditions. Formulas that followed from the fundamental decisions made at this point resulted in good results but results that were sometimes not what one might expect.

In the paper wherein he introduced quantum mechanics to the world he cautions, "A significant difficulty arises, however, if we consider two quantities $x(t)$, $y(t)$, and ask after their product....Whereas in classical $x(t)y(t)$ is always equal to $y(t)x(t)$, this is not necessarily the case in quantum theory." When the predicted values are exhibited in matrix form and multiplications are performed, the nature of the difficulty appears in a form that is more familiar to mathematicians. More significantly, empirical studies validate the theoretical results and suggest that there is something of deep importance in that the difference between x(t)y(t) and y(t)x(t) is a value related to Planck's constant.

Schema for a table of transition frequencies (produced when electrons change orbitals):

Electron StatesS1	**S2**	**S3**	**S4**	**S5**	**....**			
S1	f1→1	f2→1	f3→1	f4→1	f5→1	.....		
S2	f1→2	f2→2	f3→2	f4→2	f5→2	.....		
S3	f1→3	f2→3	f3→3	f4→3	f5→3	.....		
S4	f1→4	f2→4	f3→4	f4→4	f5→4	.....	S5	f1→5
f2→5	f3→5	f4→5	f5→5	.....	S.....	.....	.....	.....
.....	.....	.....						

Schema for a related table showing the transition amplitudes:

Electron StatesS1	***S2***	***S3***	***S4***	***S5***	**....**	
S1	a1→1	a2→1	a3→1	a4→1	a5→1	.....
S2	a1→2	a2→2	a3→2	a4→2	a5→2	.....
S3	a1→3	a2→3	a3→3	a4→3	a5→3	.....
S4	a1→4	a2→4	a3→4	a4→4	a5→4	.....
S5	a1→5	a2→5	a3→5	a4→5	a5→5	.....
S.....	.....	.....	.....	.....	.....	.....

As related above, Heisenberg developed ways of meaningfully relating the information in tables such as these in a mathematical way. Empirically filling in the values for tables involving quantum quantities is not a simple procedure since any measurement made on a single system gives that one value but has the potential of changing other values. So large numbers of identical copies of the system in question must be prepared, and a single measure made on each system. Multiple experiments to determine the same characteristics are made, and the results are averaged. Even then, precise measurements of all characteristics of the system as they would appear simultaneously cannot be provided because of quantum uncertainty.

A precise determination of one characteristic's value necessarily creates an uncertainty in the value of its correlate. "Certain pairs of observables simply *cannot* be simultaneously measured to an arbitrarily high level of precision." If simultaneous measurements are made of correlated characteristics in multiple identical systems, there will inevitably be differences in the measurements such that the difference between their products is equal to or greater than $\hbar/2$." In 1925 Heisenberg published a paper entitled "Quantum-mechanical re-interpretation of kinematic and mechanical relations" relating his discoveries.

So ended the old quantum theory and began the age of quantum mechanics. Heisenberg's paper gave few details that might aid readers in determining how he actually contrived to get his results for the one-dimensional models he used to form the hypothesis that proved so useful. In his paper, Heisenberg proposed to "discard all hope of observing hitherto unobservable quantities, such as the position and period of the electron", and restrict himself strictly to actually observable quantities. He needed mathematical rules for predicting the relations actually observed in nature, and the rules he produced worked differently depending on the sequence in which they were applied. "It quickly became clear that the *non-commutativity* of kinematical quantities in quantum theory was the really essential new technical idea in the paper."

The special type of multiplication that turned out to be required in his formula was most elegantly described by using special arrays of numbers called matrices. In ordinary situations it does not matter in which order the operations involved in multiplication are performed, but matrix multiplication does not commute. Essentially that means that it matters which order given operations are performed in. Multiplying matrix A by matrix B is not the same as multiplying matrix B by matrix A.

In symbols, A×B is in general not equal to B×A. (The important thing in quantum theory is that it turned out to matter whether one experimentally measures velocity first and then immediately measures position, or vice-versa.) The matrix convention turned out to be a convenient way of organizing information and making clear the exact sequence in which calculations must be made, and it reflects in a symbolic form the unexpected results obtained in the real world.

Heisenberg approached quantum mechanics from the historical perspective that treated an electron as an oscillating charged particle. Bohr's use of this analogy had already allowed him to explain why the radii of the orbits of electrons could only take on certain values. It followed from this interpretation of the experimental results available and the quantum theory that Heisenberg subsequently created that an electron could not be at any intermediate position between two "permitted" orbits. Therefore electrons were described as "jumping" from orbit to orbit.

The idea that an electron might now be in one place and an instant later be in some other place without having traveled between the two points was one of the earliest indications of the "spookiness" of quantum phenomena. Although the scale is smaller, the "jump" from orbit to orbit is as strange and unexpected as would be a case in which someone stepped out of a doorway in London onto the streets of Los Angeles. Quantum tunneling is one instance in which electrons seem to be able to move in the "spooky" way that Heisenberg ascribed to their actions within atoms.

Amplitudes of position and momentum that have a period of 2 π like a cycle in a wave are called Fourier series variables. Heisenberg described the particle-like properties of the electron in a wave as having position and momentum in his matrix mechanics. When these amplitudes of position and momentum are measured and multiplied together, they give intensity.

However, he found that when the position and momentum were multiplied together in that respective order, and then the momentum and position were multiplied together in that respective order, there was a difference or deviation in intensity between them of $h/2\pi$.

Heisenberg would not understand the reason for this deviation until two more years had passed, but for the time being he satisfied himself with the idea that the math worked and provided an exact description of the quantum behaviour of the electron.

Matrix mechanics was the first complete definition of quantum mechanics, its laws, and properties that described fully the behaviour of the electron. It was later extended to apply to all subatomic particles.

Very soon after matrix mechanics was introduced to the world, Schrödinger, acting independently, produced a quantum wave theory that appeared to have no similarities whatsoever to Heisenberg's theory. It was computationally easier and avoided some of the odd-sounding ideas like "quantum leaps" of an electron from one orbit to another. But within a short time Schrödinger himself had shown that the two theories produced essentially the same results in all situations. Finally, Dirac made the idea of non-commutativity central to his own theory and proved the formulations of Heisenberg and of Schrödinger to be special cases of his own theory.

Because particles could be described as waves, later in 1925 Erwin Schrödinger analysed what an electron would look like as a wave around the nucleus of the atom. Using this model, he formulated his equation for particle waves. Rather than explaining the atom by analogy to satellites in planetary orbits, he treated everything as waves whereby each electron has its own unique wavefunction. A wavefunction is described in Schrödinger's equation by three properties.

The three properties were:

- An "orbital" designation, indicating whether the particle wave is one that is closer to the nucleus with less energy or one that is farther from the nucleus with more energy,
- The shape of the orbital, *i.e.*, an indication that orbitals were not just spherical but other shapes, and
- The inclination of the orbital, which determines the magnetic moment of the orbital around the z-axis.

These three properties were called collectively the wavefunction of the electron and are said to describe the quantum state of the electron. "Quantum state" means the collective properties of the electron describing what we can say about its condition at a given time. For the electron, the quantum state is described by its wavefunction, which is designated in physics by the Greek letter ψ (psi, pronounced "sigh"). The three properties of Schrödinger's equation that describe the wavefunction of the electron and, therefore, also describe the quantum state of the electron as described in the previous paragraph are each called quantum numbers. The first property that describes the orbital was numbered according to Bohr's model where n is the letter used to describe the energy of each orbital. This is called the principal quantum number.

The next quantum number that describes the shape of the orbital is called the azimuthal quantum number and it is represented by the letter l (lower case L). The shape is caused by the angular momentum of the orbital. The rate of change of the angular momentum of any system is equal to the resultant external torque acting on that system. In other words, angular momentum represents the resistance of a spinning object to speed up or slow down under the influence of external force.

The azimuthal quantum number "l" represents the orbital angular momentum of an electron around its nucleus. However, the shape of each orbital has its own letter as well. So for the letter "l" there are other letters to describe the shapes of "l". The first shape is spherical and is described by the letter s. The next shape is like a dumbbell and is described by the letter p. The other shapes of orbitals become more complicated and are described by the letters d, f, and g. For the shape of a carbon atom. The third quantum number of Schrödinger's equation describes the magnetic moment of the electron and is designated by the letter m and sometimes as the letter m with a subscript l because the magnetic moment depends upon the second quantum number l.

In May 1926 Schrödinger published a proof that Heisenberg's matrix mechanics and his own wave mechanics gave equivalent results: mathematically they were the same theory. Yet both men disagreed on the interpretation of this theory. Heisenberg saw no problem in the existence of discontinuous quantum jumps, while Schrödinger hoped that a theory based on continuous wave-like properties could avoid this "nonsense about quantum jumps".

Uncertainty Principle

In 1927, Heisenberg made a new discovery on the basis of his quantum theory that had further practical consequences of this new way of looking at matter and energy on the atomic scale. In Heisenberg's matrix mechanics formula, he encountered an error or difference of $h/2\pi$ between position and momentum.

The more certain the position of a particle is determined, the less certain the momentum is known, and $h/2\pi$ is the lower limit of the uncertainty involved. This conclusion came to be called "Heisenberg's Indeterminacy Principle", or Heisenberg's Uncertainty Principle.For moving particles in quantum mechanics, there is simply a certain degree of exactness and precision that is missing. The observer can be precise when taking a measurement of position or can be precise when taking a measurement of momentum, but there is an inverse imprecision when measuring both at the same time as in the case of a moving particle like the electron. One consequence of the uncertainty principle was that the electron could no longer be considered as in an exact location in its orbital.

Rather the electron had to be described by every point where the electron could possibly inhabit. Calculating points of probable location for the electron in its known orbital created the picture of a cloud of points in a spherical shape for the orbital of a hydrogen atom which points gradually faded out nearer to the nucleus and farther from the nucleus. This picture may be termed a probability distribution. Therefore, the Bohr atom number *n* for each orbital became known as an *n-sphere* in the three dimensional atom and was pictured as a probability cloud where the electron surrounded the atom all at once.

This led to the further description by Heisenberg that if a measurement of the electron was not being taken that it could not be described in one particular location but was everywhere in the electron cloud at once. In other words, quantum mechanics cannot give exact results, but only the probabilities for the occurrence of a variety of possible results.

Heisenberg went further and said that the path of a moving particle only comes into existence once we observe it. However strange and counter-intuitive this assertion may seem, quantum mechanics does still tell us the location of the electron's orbital, its probability cloud. Heisenberg was speaking of the particle itself, not its orbital which is in a known probability distribution.

It is important to note that although Heisenberg used infinite sets of positions for the electron in his matrices, this does not mean that the electron could be anywhere in the universe. Rather there are several laws that show the electron must be in one localized probability distribution. An electron is described by its energy in Bohr's atom which was carried over to matrix mechanics.

Therefore, an electron in a certain n-sphere had to be within a certain range from the nucleus depending upon its energy. This restricts its location. Also, the number of places an electron can be is also called "the number of cells in its phase space". The Uncertainty Principle set a lower limit to how finely one can chop up classical phase space, so the number of places that an electron can be in its orbital becomes finite. An electron's location in an atom is defined to be in its orbital, but stops at the nucleus and before the next n-sphere orbital begins.

Classical physics had shown since Newton that if the position of stars and planets and details about their motions were known then where they will be in the future can be predicted.

For subatomic particles, Heisenberg denied this notion showing that due to the uncertainty principle one cannot know the precise position and momentum of a particle at a given instant, so its future motion cannot be determined, but only a range of possibilities for the future motion of the particle can be described.

These notions arising from the uncertainty principle only arise at the subatomic level and were a consequence of wave-particle duality. As counter-intuitive as they may seem, quantum mechanical theory with its uncertainty principle has been responsible for major improvements in the world's technology from computer components to fluorescent lights to brain scanning techniques.

Wavefunction Collapse

Schrödinger's wave equation with its unique wavefunction for a single electron is also spread out in a probability distribution like Heisenberg's quantized particle-like electron. This is because a wave is naturally a widespread disturbance and not a point particle. Therefore, Schrödinger's wave equation has the same predictions made by the uncertainty principle because uncertainty of location is built into the definition of a widespread disturbance like a wave.

Uncertainty only needed to be defined from Heisenberg's matrix mechanics because the treatment was from the particle-like aspects of the electron. Schrödinger's wave equation shows that the electron is in the probability cloud at all times in its probability distribution as a wave that is spread out. Max Born discovered in 1928 that when the square of Schrödinger's wavefunction (psi-squared) is computed the electron's location as a probability distribution is obtained.

Therefore, to the extent that a measurement of the position of an electron can be made at an exact location instead of as a probability distribution, the electron appears to momentarily cease to have wave-like properties. Without wave-like properties, none of Schrödinger's definitions of the electron being wave-like makes sense. The measurement of the position of the particle nullifies the simple wave-like properties and the one body form of Schrödinger's equation then fails. Because the electron can no longer be described by its separate wavefunction when measured, due to its wave length becoming much shorter and its becoming entangled with the particles of the measuring apparatus, this is called wavefunction collapse.

The term eigenstate is derived from the German/Dutch word *"eigen,"* which means "inherent" or "characteristic." The word eigenstate is descriptive of the measured state of some object that possesses quantifiable characteristics such as position, momentum, etc. The state being measured and described must be an "observable" (*i.e.* something that can be experimentally measured either directly or indirectly like position or momentum), and must have a definite value.

In the everyday world, it is natural and intuitive to think of everything being in its own eigenstate. Everything appears to have a definite position, a definite momentum, a definite value of measure, and a definite time of occurrence. However, quantum mechanics affirms that it is impossible to pinpoint exact values for the momentum of a certain particle like an electron in a given location at a particular moment in time, or, alternatively, that it is impossible to give an exact location for such an object when the momentum has been measured. Due to the uncertainty principle, statements regarding both the position and momentum of particles can only be given in terms of a range of probabilities, a "probability distribution". Eliminating uncertainty in one term maximizes uncertainty in regard to the second parameter.

Therefore it became necessary to have a way to clearly formulate the difference between the state of something that is uncertain in the way just described, such as an electron in a probability cloud, and effectively contrast it to the state of something that is not uncertain, something that has a definite value. When something is in the condition of being definitely "pinned-down" in some regard, it is said to possess an eigenstate. For example, if the position of an electron has been made definite, it is said to have an eigenstate of position.

A definite value, such as the position of an electron that has been successfully located, is called the eigenvalue of the eigenstate of position. The German word *"eigen"* was first used in this context by the mathematician David Hilbert in 1904. Schrödinger's wave equation gives wavefunction solutions, meaning the possibilities where the electron might be, just as does Heisenberg's probability distribution. As stated above, when a wavefunction collapse occurs because something has been done to locate the position of an electron, the electron's state becomes an eigenstate of position, meaning that the position has a known value. The Pauli Exclusion Principle states that no electron (or other fermion) can be in the same quantum state as another within an atom. Wolfgang Pauli developed the Exclusion Principle from what he called a "two-valued quantum degree of freedom" to account for the observation of a doublet, meaning a pair of lines differing by a small amount (*e.g.*, something on the order of 0.15Å), in the spectrum of the hydrogen atom.

The observation meant that there was more energy in the electron orbital from magnetic moment than had previously been described. In early 1925, the young physicists Uhlenbeck and Goudsmit introduced a theory that the electron rotates in space in the same way that the earth rotates on its axis.

This would account for the missing magnetic moment and allow for two electrons in the same orbital to be different if their spins were in opposite directions, thus satisfying the Exclusion Principle. A new quantum number was then needed, one to represent the momentum involved in the rotation of each electron. One concise statement of Pauli's principle is as follows:

There cannot exist an atom in such a quantum state that two electrons within have the same set of quantum numbers.

An electron has four quantum numbers:

- n, the total quantum number of the orbit
- l, the azimuthal quantum number
- m_l, the orbital magnetic quantum number
- m_s, the spin magnetic quantum number

"In the case of a helium atom with two electrons in the 1*s* orbital, the Pauli exclusion principle requires that the two electrons differ in the value of one quantum number. Their values of n, l, and m_l are the same; moreover, they have the same spin quantum number, s = 1/2. Accordingly they must differ in the value of m_s, which can have the value of +½ for one electron and –½ for the other."

Dirac Wave Equation: In 1928, Paul Dirac extended the Pauli equation, which described spinning electrons, to account for relativity. By using the simplest electromagnetic interaction, he was able to predict the value of the magnetic moment associated with the electron's spin, and found the experimentally observed value, a value too large to be that of a spinning classical charged sphere. He was able to solve for the spectral lines of the Hydrogen atom, reproducing Sommerfeld's successful formula for the fine-structure splitting of the spectral lines from sound physical principles.

Dirac's theory suffered from a problem of negative energy, which required Dirac to introduce antiparticles and a dynamical vacuum and led to many-particle quantum field theory. All of the above development of quantum theory was based mainly on the atomic spectrum of the hydrogen atom. This is due to the fact that each atom of each element produces a unique pattern of spectral lines when light from each different kind of element is passed through a prism. Scientists could not study the electron and nucleus of the atom itself because they cannot be seen.

Even today with High-resolution Scanning Tunneling Electron Microscopes we can only get images of the atom as a blurry fuzzball. However, the spectral lines of the atom reveal the orbits of electrons and the energy that can be expected. It was basically this study of the spectroscopic analysis of first the hydrogen atom and then the helium atom that led to quantum theory.

Therefore, the mathematical formulas were made to fit the picture of the atomic spectrum. That is why quantum mechanics is sometimes referred to as a form of mathematical physics.

Quantum Entanglement

Albert Einstein rejected Heisenberg's Uncertainty Principle insofar as it seemed to imply more than a necessary limitation on human ability to actually know what occurs in the quantum realm. In a letter to Max Born in 1926, Einstein famously declared that "God does not play dice". The bare surface level prescription for making predictions from quantum mechanics, based on Born's rule for computing probabilities, became known as the Copenhagen Interpretation of quantum mechanics. Bohr spent many years developing and refining this interpretation in light of Einstein's objections. After the 1930 Solvay conference, Einstein never again challenged the Copenhagen interpretation on technical grounds, but did not cease a philosophical attack on the interpretation, on the grounds of realism and locality (*i.e.*, the idea that there is no action at a distance).

Einstein, in trying to show that quantum theory was not a complete theory, recognized that the theory predicted that two or more particles that had interacted in the past can exhibit strong correlations later on when various measurements are made on them. He wanted this seeming interaction to be explained in a classical way through their common past, and preferably not by a "spooky action at a distance". The argument is worked out in a famous paper devoted to what is now called the EPR paradox. Assuming what is now usually called "local realism", the EPR attempts to show from quantum theory that particles simultaneously possess both position and momentum, while according to the Copenhagen interpretation, only one of these two properties actually exists, and only in the moment that it is being measured.

Einstein considered this conclusion a proof that quantum theory is incomplete in that it refuses to physical properties which objectively exist in nature. The feature of quantum theory leading to these paradoxes is called quantum entanglement. Quantum entanglement refers to situations wherein the properties of several discrete objects cannot be described simply by considering them separately, even after taking account of the history of their past interaction. A quantum interaction can affect the entangled quantum entity in a like manner even though the two of them are widely separated. Bohr's original response to Einstein was that entangled particles are part of one indivisible system.

Einstein's challenge led to decades of substantial research into quantum entanglement. Research would seem to confirm Bohr's assertion that two or more entangled particles must be viewed together as one whole, and, moreover, that difficulties only arise by insisting on the reality of outcomes of measurements that are not made anyway. It might be said that God does indeed throw dice, though rather peculiar ones. A real dice throw can be completely understood using the laws of classical mechanics, and the outcome is merely a function of the initial conditions. However the outcome of tossing quantum "dice" has no antecedent; it has as yet no discernible cause.

Interpretations

Despite the success of quantum mechanics, it does have some controversial elements. For example, the behaviour of microscopic objects described in quantum mechanics is very different from our everyday experience, which may provoke some degree of incredulity. Most of classical physics is now recognized to be composed of special cases of quantum physics theory and/or relativity theory.

Dirac brought relativity theory to bear on quantum physics so that it could properly deal with events that occur at a substantial fraction of the speed of light. Classical physics, however, also deals with mass attraction (gravity), and no one has yet been able to bring gravity into a unified theory with the relativized quantum theory.Also interesting, according to the correspondence principle and Ehrenfest's theorem as a system becomes larger or more massive (action >> h) the classical dynamics tends to emerge (with some exceptions such as superfluidity).

This is why we can usually ignore quantum mechanics when dealing with everyday objects; instead the classical description will suffice. Even so, trying to make sense of quantum theory is an ongoing process that has spawned a number of interpretations of quantum theory, ranging from the conventional Copenhagen Interpretation to hidden variables and many worlds interpretations. There seems to be no end in sight to the philosophical musings on the subject; however the empirical or technical success of the theory is unrivaled; all modern fundamental physical theories are quantum theories, relativity being subsumed within quantum field theories.

Relativity and Quantum Mechanics

The modern world of physics is founded on two tested and demonstrably sound theories of general relativity and quantum mechanics —theories which appear to contradict one another. The defining postulates of both Einstein's theory of relativity and quantum theory are indisputably supported by rigorous and repeated empirical evidence. However, while they do not directly contradict each other theoretically they are resistant to being incorporated within one cohesive model.Einstein himself is well known for rejecting some of the claims of quantum mechanics. While clearly inventive in this field, he did not accept the more philosophical consequences and interpretations of quantum mechanics, such as the lack of deterministic causality and the assertion that a single subatomic particle can occupy numerous areas of space at one time.

He also was the first to notice some of the apparently exotic consequences of entanglement and used them to formulate the Einstein-Podolsky-Rosen paradox, in the hope of showing that quantum mechanics has unacceptable implications. This was 1935, but in 1964 it was shown by John Bell that Einstein's assumption that quantum mechanics is correct, but has to be completed by *hidden variables*, was based on wrong philosophical assumptions: according to the paper of J. Bell and the Copenhagen interpretation (the common interpretation of quantum mechanics by physicists for decades), and contrary to Einstein's ideas, quantum mechanics is:

- Neither a "realistic" theory (since quantum measurements do not *state* pre-existing properties, but rather they *prepare* properties)
- Nor a *local* theory (essentially not, because the state vector $|\psi\rangle$ determines simultaneously the probability amplitudes at all sites, $|\psi\rangle \rightarrow \psi(r)$, $\forall$r).

The Einstein-Podolsky-Rosen paradox shows in any case that there exist experiments by which one can measure the state of one particle and instantaneously change the state of its entangled partner, although the two particles can be an arbitrary distance apart; however, this effect does not violate causality, since no transfer of information happens. These experiments are the basis of some of the most topical applications of the theory, quantum cryptography, which works well, although at small distances of typically ≤ 1000 km, being on the market since 2004.

There do exist quantum theories which incorporate *special* relativity-for example, quantum electrodynamics (QED), which is currently the most accurately tested physical theory-and these lie at the very heart of modern particle physics. Gravity is negligible in many areas of particle physics, so that unification between general relativity and quantum mechanics is not an urgent issue in those applications. However, the lack of a correct theory of quantum gravity is an important issue in cosmology.

Attempts at a Unified Theory

Inconsistencies arise when one tries to join the quantum laws with general relativity, a more elaborate description of spacetime which incorporates gravitation. Resolving these inconsistencies has been a major goal of twentieth- and twenty-first-century physics. Many prominent physicists, including Stephen Hawking, have labored in the attempt to discover a "Grand Unification Theory" that combines not only different models of subatomic physics, but also derives the universe's four forces—the strong force, electromagnetism, weak force, and gravity— from a single force or phenomenon.

Quantum Mechanics and Classical Physics

Predictions of quantum mechanics have been verified experimentally to a very high degree of accuracy. Thus, the current logic of correspondence principle between classical and quantum mechanics is that all objects obey laws of quantum mechanics, and classical mechanics is just a quantum mechanics of large systems (or a statistical quantum mechanics of a large collection of particles).

Laws of classical mechanics thus follow from laws of quantum mechanics at the limit of large systems or large quantum numbers.Main differences between classical and quantum theories have already been mentioned above in the remarks on the Einstein-Podolsky-Rosen paradox. Essentially the difference boils down to the statement that quantum mechanics is coherent (addition of *amplitudes*), whereas classical theories are incoherent (addition of *intensities*). Thus, such quantities as *coherence lengths* and *coherence times* come into play. For microscopic bodies the extension of the system is certainly much smaller than the coherence length; for macroscopic bodies one expects that it should be the other way round.Many "macroscopic" properties of "classic" systems are direct consequences of quantum behaviour of its parts.

For example, stability of bulk matter (which consists of atoms and molecules which would quickly collapse under electric forces alone), rigidity of this matter, mechanical, thermal, chemical, optical and magnetic properties of this matter—they are all results of interaction of electric charges under the rules of quantum mechanics. Whilst seemingly exotic behaviour of matter posited by quantum mechanics and relativity theory become more apparent when dealing with extremely fast-moving or extremely tiny particles, the laws of classical "Newtonian" physics still remain accurate in predicting the behaviour of surrounding ("large") objects—of the order of the size of large molecules and bigger—at velocities much smaller than the velocity of light. There are numerous mathematically equivalent formulations of quantum mechanics.

One of the oldest and most commonly used formulations is the transformation theory proposed by Cambridge theoretical physicist Paul Dirac, which unifies and generalizes the two earliest formulations of quantum mechanics, matrix mechanics (invented by Werner Heisenberg) and wave mechanics. In this formulation, the instantaneous state of a quantum system encodes the probabilities of its measurable properties, or "observables". Examples of observables include energy, position, momentum, and angular momentum. Observables can be either continuous (*e.g.*, the position of a particle) or discrete (*e.g.*, the energy of an electron bound to a hydrogen atom). Generally, quantum mechanics does not assign definite values to observables.

Instead, it makes predictions about probability distributions; that is, the probability of obtaining each of the possible outcomes from measuring an observable. Naturally, these probabilities will depend on the quantum state at the instant of the measurement. There are, however, certain states that are associated with a definite value of a particular observable. These are known as "eigenstates" of the observable ("eigen" can be roughly translated from German as inherent or as a characteristic). In the everyday world, it is natural and intuitive to think of everything being in an eigenstate of every observable. Everything appears to have a definite position, a definite momentum, and a definite time of occurrence.

However, quantum mechanics does not pinpoint the exact values for the position or momentum of a certain particle in a given space in a finite time; rather, it only provides a range of probabilities of where that particle might be. Therefore, it became necessary to use different words.for the state of something having an *uncertainty relation* and a state that has a *definite* value.

The latter is called the "eigenstate" of the property being measured. For example, consider a free particle. In quantum mechanics, there is wave-particle duality so the properties of the particle can be described as a wave. Therefore, its quantum state can be represented as a wave, of arbitrary shape and extending over all of space, called a wave function. The position and momentum of the particle are observables. The Uncertainty Principle of quantum mechanics states that both the position and the momentum cannot simultaneously be known with infinite precision at the same time.

However, one can measure just the position alone of a moving free particle creating an eigenstate of position with a wavefunction that is very large at a particular position x, and almost zero everywhere else. If one performs a position measurement on such a wavefunction, the result x will be obtained with almost 100% probability. In other words, the position of the free particle will almost be known.

This is called an eigenstate of position (mathematically more precise: a *generalized eigenstate (eigendistribution)*). If the particle is in an eigenstate of position then its momentum is completely unknown. An eigenstate of momentum, on the other hand, has the form of a plane wave. It can be shown that the wavelength is equal to h/p, where h is Planck's constant and p is the momentum of the eigenstate. If the particle is in an eigenstate of momentum then its position is completely blurred out.

Usually, a system will not be in an eigenstate of whatever observable we are interested in. However, if one measures the observable, the wavefunction will instantaneously be an eigenstate of that observable. This process is known as wavefunction collapse. It involves expanding the system under study to include the measurement device, so that a detailed quantum calculation would no longer be feasible and a classical description must be used. If one knows the corresponding wave function at the instant before the measurement, one will be able to compute the probability of collapsing into each of the possible eigenstates.

For example, the free particle in the previous example will usually have a wavefunction that is a wave packet centered around some mean position x_0, neither an eigenstate of position nor of momentum. When one measures the position of the particle, it is impossible to predict with certainty the result that we will obtain. It is probable, but not certain, that it will be near x_0, where the amplitude of the wave function is large. After the measurement is performed, having obtained some result x, the wave function collapses into a position eigenstate centered at x.

Wave functions can change as time progresses. An equation known as the Schrödinger equation describes how wave functions change in time, a role similar to Newton's second law in classical mechanics. The Schrödinger equation, applied to the aforementioned example of the free particle, predicts that the centre of a wave packet will move through space at a constant velocity, like a classical particle with no forces acting on it.

However, the wave packet will also spread out as time progresses, which means that the position becomes more uncertain. This also has the effect of turning position eigenstates into broadened wave packets that are no longer position eigenstates.Some wave functions produce probability distributions that are constant in time. Many systems that are treated dynamically in classical mechanics are described by such "static" wave functions.

For example, a single electron in an unexcited atom is pictured classically as a particle moving in a circular trajectory around the atomic nucleus, whereas in quantum mechanics it is described by a static, spherically symmetric wavefunction surrounding the nucleus. The time evolution of wave functions is deterministic in the sense that, given a wavefunction at an initial time, it makes a definite prediction of what the wavefunction will be at any later time.

During a measurement, the change of the wavefunction into another one is not deterministic, but rather unpredictable, *i.e.*, random.The probabilistic nature of quantum mechanics thus stems from the act of measurement. This is one of the most difficult aspects of quantum systems to understand. It was the central topic in the famous Bohr-Einstein debates, in which the two scientists attempted to clarify these fundamental principles by way of thought experiments. In the decades after the formulation of quantum mechanics, the question of what constitutes a "measurement" has been extensively studied. Interpretations of quantum mechanics have been formulated to do away with the concept of "wavefunction collapse" the relative state interpretation. The basic idea is that when a quantum system interacts with a measuring apparatus, their respective wavefunctions become entangled, so that the original quantum system ceases to exist as an independent entity.

Mathematical Formulation

In the mathematically rigorous formulation of quantum mechanics, developed by Paul Dirac and John von Neumann, the possible states of a quantum mechanical system are represented by unit vectors (called "state vectors") residing in a complex separable Hilbert space well defined up to a complex number of norm 1 (the phase factor). In other words, the possible states are points in the projectivization of a Hilbert space, usually called the complex projective space. The exact nature of this Hilbert space is dependent on the system; for example, the state space for position and momentum states is the space of square-integrable functions, while the state space for the spin of a single proton is just the product of two complex planes.

Each observable is represented by a maximally-Hermitian linear operator acting on the state space. Each eigenstate of an observable corresponds to an eigenvector of the operator, and the associated eigenvalue corresponds to the value of the observable in that eigenstate. If the operator's spectrum is discrete, the observable can only attain those discrete eigenvalues.The time evolution of a quantum state is described by the Schrödinger equation, in which the Hamiltonian, the operator corresponding to the total energy of the system, generates time evolution.

The inner product between two state vectors is a complex number known as a probability amplitude. During a measurement, the probability that a system collapses from a given initial state to a particular eigenstate is given by the square of the absolute value of the probability amplitudes between the initial and final states. The possible results of a measurement are the eigenvalues of the operator-which explains the choice of *Hermitian* operators, for which all the eigenvalues are real.

We can find the probability distribution of an observable in a given state by computing the spectral decomposition of the corresponding operator. Heisenberg's uncertainty principle is represented by the statement that the operators corresponding to certain observables do not commute.The Schrödinger equation acts on the entire probability amplitude, not merely its absolute value.

Whereas the absolute value of the probability amplitude encodes information about probabilities, its phase encodes information about the interference between quantum states. This gives rise to the wave-like behaviour of quantum states.

It turns out that analytic solutions of Schrödinger's equation are only available for a small number of model Hamiltonians, of which the quantum harmonic oscillator, the particle in a box, the hydrogen molecular ion and the hydrogen atom are the most important representatives.

Even the helium atom, which contains just one more electron than hydrogen, defies all attempts at a fully analytic treatment. There exist several techniques for generating approximate solutions. For instance, in the method known as perturbation theory one uses the analytic results for a simple quantum mechanical model to generate results for a more complicated model related to the simple model by, for example, the addition of a weak potential energy.

Another method is the "semi-classical equation of motion" approach, which applies to systems for which quantum mechanics produces weak deviations from classical behaviour. The deviations can be calculated based on the classical motion. This approach is important for the field of quantum chaos.An alternative formulation of quantum mechanics is Feynman's path integral formulation, in which a quantum-mechanical amplitude is considered as a sum over histories between initial and final states; this is the quantum-mechanical counterpart of action principles in classical mechanics.

Interactions with Other Scientific Theories

The fundamental rules of quantum mechanics are very broad. They assert that the state space of a system is a Hilbert space and the observables are Hermitian operators acting on that space, but do not tell us which Hilbert space or which operators, or if it even exists. These must be chosen appropriately in order to obtain a quantitative description of a quantum system. An important guide for making these choices is the correspondence principle, which states that the predictions of quantum mechanics reduce to those of classical physics when a system moves to higher energies or equivalently, larger quantum numbers.

In other words, classic mechanics is simply a quantum mechanics of large systems. This "high energy" limit is known as the *classical* or *correspondence limit*. One can therefore start from an established classical model of a particular system, and attempt to guess the underlying quantum model that gives rise to the classical model in the correspondence limit.

When quantum mechanics was originally formulated, it was applied to models whose correspondence limit was non-relativistic classical mechanics. For instance, the well-known model of the quantum harmonic oscillator uses an explicitly non-relativistic expression for the kinetic energy of the oscillator, and is thus a quantum version of the classical harmonic oscillator.Early attempts to merge quantum mechanics with special relativity involved the replacement of the Schrödinger equation with a covariant equation such as the Klein-Gordon equation or the Dirac equation. While these theories were successful in explaining many experimental results, they had certain unsatisfactory qualities stemming from their neglect of the relativistic creation and annihilation of particles.

A fully relativistic quantum theory required the development of quantum field theory, which applies quantization to a field rather than a fixed set of particles. The first complete quantum field theory, quantum electro-dynamics, provides a fully quantum description of the electromagnetic interaction.

The full apparatus of quantum field theory is often unnecessary for describing electrodynamic systems. A simpler approach, one employed since the inception of quantum mechanics, is to treat charged particles as quantum mechanical objects being acted on by a classical electromagnetic field. For example, the elementary quantum model of the hydrogen atom describes the electric field of the hydrogen atom using a classical,

$$-\frac{e^2}{4\pi\epsilon_0}\frac{1}{r}$$

Coulomb potential. This "semi-classical" approach fails if quantum fluctuations in the electromagnetic field play an important role, such as in the emission of photons by charged particles. Quantum field theories for the strong nuclear force and the weak nuclear force have been developed.

The quantum field theory of the strong nuclear force is called quantum chromodynamics, and describes the interactions of the subnuclear particles: quarks and gluons.

The weak nuclear force and the electromagnetic force were unified, in their quantized forms, into a single quantum field theory known as electroweak theory, by the physicists Abdus Salam, Sheldon Glashow and Steven Weinberg. It has proven difficult to construct quantum models of gravity, the remaining fundamental force.

Semi-classical approximations are workable, and have led to predictions such as Hawking radiation. However, the formulation of a complete theory of quantum gravity is hindered by apparent incompatibilities between general relativity, the most accurate theory of gravity currently known, and some of the fundamental assumptions of quantum theory.

The resolution of these incompatibilities is an area of active research, and theories such as string theory are among the possible candidates for a future theory of quantum gravity.

Derivation of Quantization

The particle in a 1-dimensional potential energy box is the most simple example where restraints lead to the quantization of energy levels. The box is defined as zero potential energy inside a certain interval and infinite everywhere outside that interval. For the 1-dimensional case in the x direction, the time-independent Schrödinger equation can be written as:

$$= \psi.$$

The general solutions are:

$$y = Ae^{ikx} + Be^{-ikx} \; E =$$

$$y = C \sin kx + D \cos kx \text{ (exponential rewrite)}$$

The presence of the walls of the box restricts the acceptable solutions to the wavefunction.

At each wall:

$$\psi = 0 \text{ at } x = 0,\ x = L$$

Consider $x = 0$

- Sin 0 = 0, cos 0 = 1. To satisfy $\psi = 0$ $D = 0$ (cos term is removed)
- *Now Consider*: $\psi = C \sin kx$ at $X = L$, $\psi = C \sin kL$

- If $C = 0$ then for all x and would conflict with Born interpretation
- Therefore sin kL.

In this situation, n must be an integer showing the quantization of the energy levels.

Applications

Quantum mechanics has had enormous success in explaining many of the features of our world. The individual behaviour of the subatomic particles that make up all forms of matter—electrons, protons, neutrons, photons and others—can often only be satisfactorily described using quantum mechanics. Quantum mechanics has strongly influenced string theory, a candidate for a theory of everything and the multiverse hypothesis. It is also related to statistical mechanics.

Quantum mechanics is important for understanding how individual atoms combine covalently to form chemicals or molecules. The application of quantum mechanics to chemistry is known as quantum chemistry. Quantum mechanics can in principle mathematically describe most of chemistry. Quantum mechanics can provide quantitative insight into ionic and covalent bonding processes by explicitly showing which molecules are energetically favourable to which others, and by approximately how much. Most of the calculations performed in computational chemistry rely on quantum mechanics.Much of modern technology operates at a scale where quantum effects are significant. Examples include the laser, the transistor, the electron microscope, and magnetic resonance imaging. The study of semiconductors led to the invention of the diode and the transistor, which are indispensable for modern electronics. Researchers are currently seeking robust methods of directly manipulating quantum states.

Efforts are being made to develop quantum cryptography, which will allow guaranteed secure transmission of information. A more distant goal is the development of quantum computers, which are expected to perform certain computational tasks exponentially faster than classical computers.

Another active research topic is quantum teleportation, which deals with techniques to transmit quantum states over arbitrary distances.In many devices, even the simple light switch, quantum tunneling is vital, as otherwise the electrons in the electric current could not penetrate the potential barrier made up, in the case of the light switch, of a layer of oxide. Flash memory chips found in USB drives also use quantum tunneling to erase their memory cells.

Philosophical Consequences

Since its inception, the many counter-intuitive results of quantum mechanics have provoked strong philosophical debate and many interpretations. Even fundamental issues such as Max Born's basic rules concerning probability amplitudes and probability distributions took decades to be appreciated. The Copenhagen interpretation, due largely to the Danish theoretical physicist Niels Bohr, is the interpretation of quantum mechanics most widely accepted amongst physicists. According to it, the probabilistic nature of quantum mechanics predictions cannot be explained in terms of some other deterministic theory, and does not simply reflect our limited knowledge. Quantum mechanics provides probabilistic results because the physical universe is itself probabilistic rather than deterministic.

Albert Einstein, himself one of the founders of quantum theory, disliked this loss of determinism in measurement. He held that there should be a local hidden variable theory underlying quantum mechanics and consequently the present theory was incomplete.

He produced a series of objections to the theory, the most famous of which has become known as the EPR paradox. John Bell showed that the EPR paradox led to experimentally testable differences between quantum mechanics and local realistic theories. Experiments have been performed confirming the accuracy of quantum mechanics, thus demonstrating that the physical world cannot be described by local realistic theories.The writer C. S. Lewis viewed quantum mechanics as incomplete. Lewis, a professor of English, was of the opinion that the Heisenberg uncertainty principle was more of an epistemic limitation than an indication of ontological indeterminacy, and in this respect believed similarly to many advocates of hidden variables theories.

The *Bohr-Einstein debates* provide a vibrant critique of the Copenhagen Interpretation from an epistemological point of view.The Everett many-worlds interpretation, formulated in 1956, holds that all the possibilities described by quantum theory simultaneously occur in a "multiverse" composed of mostly independent parallel universes.

This is not accomplished by introducing some new axiom to quantum mechanics, but on the contrary by *removing* the axiom of the collapse of the wave packet: All the possible consistent states of the measured system and the measuring apparatus are present in a *real* physical quantum superposition. While the multiverse is deterministic, we perceive non-deterministic behaviour governed by probabilities, because we can observe only the universe, *i.e.* the consistent state contribution to the mentioned superposition, we inhabit.

Everett's interpretation is perfectly consistent with John Bell's experiments and makes them intuitively understandable. However, according to the theory of quantum decoherence, the parallel universes will never be accessible to us. This inaccessibility can be understood as follows: once a measurement is done, the measured system becomes entangled with both the physicist who measured it and a huge number of other particles, some of which are photons flying away towards the other end of the universe; in order to prove that the wave function did not collapse one would have to bring all these particles back and measure them again, together with the system that was measured originally. This is completely impractical, but even if one can theoretically do this, it would destroy any evidence that the original measurement took place.

Chapter 5

Atomic Structure of Elements

Atomic Nucleus

The nucleus is the very dense region consisting of nucleons (protons and neutrons) at the center of an atom. Almost all of the mass in an atom is made up from the protons and neutrons in the nucleus, with a very small contribution from the orbiting electrons. It was discovered in 1911, as a result of Ernest Rutherford's interpretation of the famous 1909 Rutherford experiment performed by Hans Geiger and Ernest Marsden, under the direction of Rutherford. The proton-neutron model of nucleus was proposed by Dmitry Ivanenko in 1932. The diameter of the nucleus is in the range of 1.75 fm (1.75×10^{-15} m) for hydrogen (the diameter of a single proton) to about 15 fm for the heaviest atoms, such as uranium. These dimensions are much smaller than the diameter of the atom itself (nucleus + electronic cloud), by a factor of about 23,000 (uranium) to about 145,000 (hydrogen). The branch of physics concerned with studying and understanding the atomic nucleus, including its composition and the forces which bind it together, is called nuclear physics.

Etymology

The term nucleus is from the Latin word *nucleus* ("nut"). In 1844, Michael Faraday used the term to refer to the "central point of an atom". The modern atomic meaning was proposed by Ernest Rutherford in 1912. The adoption of the term "nucleus" to atomic theory, however, was not immediate. In 1916, for example, Gilbert N. Lewis stated, in his famous article *The Atom and the Molecule*, that "the atom is composed of the *kernel* and an outer atom or *shell*"

Nuclear Makeup

The nucleus of an atom consists of protons and neutrons (two types of baryons) bound by the nuclear force (also known as the residual strong force). These baryons are further composed of subatomic fundamental particles known as quarks bond by the strong interaction. Which chemical element an atom represents is determined by the number of protons in the nucleus. Each proton carries a single positive charge, and the total electrical charge of the nucleus is spread fairly uniformly throughout its body, with a fall-off at the edge. Major exceptions to this rule are the light elements hydrogen and helium, where the charge is concentrated most highly at the single central point (without a volume of uniform charge), as would be expected for fermions (in this case, protons) in 1s states without orbital angular momentum. As each proton carries a unit of charge, the charge distribution is indicative of the proton distribution. The neutron distribution probably is similar.

Protons and Neutrons

Protons and neutrons are fermions, with different values of the isospin quantum number, so two protons and two neutrons can share the same space wave function since they are not identical quantum entities. They sometimes are viewed as two different quantum states of the same particle, the *nucleon*. Two fermions, such as two protons, or two neutrons, or a proton + neutron (the deuteron) can exhibit bosonic behaviour when they become loosely bound in pairs. In the rare case of a hypernucleus, a third baryon called a hyperon, with a different value of the strangeness quantum number can also share the wave function. However, the latter type of nuclei are extremely unstable and are not found on Earth except in high energy physics experiments. The neutron has a positively charged core of radius H" 0.3 fm surrounded by a compensating negative charge of radius between 0.3 fm and 2 fm. The proton has an approximately exponentially decaying positive charge distribution with a mean square radius of about 0.8 fm.

Forces

Nuclei are bound together by the residual strong force (nuclear force). The residual strong force is minor residuum of the strong interaction which binds quarks together to form protons and neutrons. This force is much weaker *between* neutrons and protons because it is mostly neutralized within them, in the same way that electromagnetic forces *between* neutral atoms (such as van der Waals forces that act between two inert gas atoms) are much weaker than the electromagnetic forces that hold the parts of the atoms internally together (for example, the forces that hold the electrons in an inert gas atom bound to its nucleus).

The nuclear force is highly attractive at the distance of typical nucleon separation, and this overwhelms the repulsion between protons which is due to the electromagnetic force, thus allowing nuclei to exist. However, because the residual strong force has a limited range because it decays quickly with distance, only nuclei smaller than a certain size can be completely stable. The largest known completely stable (e.g., stable to alpha, beta, and gamma decay) nucleus is lead-208 which contains a total of 208 nucleons (126 neutrons and 82 protons). Nuclei larger than this maximal size of 208 particles are unstable and (as a trend) become increasingly short-lived with larger size, as the number of neutrons and protons which compose them increases beyond this number. However, bismuth-209 is also stable to beta decay and has the longest half-life to alpha decay of any known isotope, estimated at a billion times longer than the age of the universe. The residual strong force is effective over a very short range (usually only a few fermis; roughly one or two nucleon diameters) and causes an attraction between any pair of nucleons. For example, between protons and neutrons to form [NP] deuteron, and also between protons and protons, and neutrons and neutrons. It also is effective for the stability of one 3-body nucleon system [PNP], helium-3, while the triton [NPN] is unstable and decays to helium-3.

Halo Nuclei and Strong Force Range Limits

The effective absolute limit of the range of the strong force is represented by halo nuclei such as lithium-11 or boron-14, in which dineutrons, or other collections of neutrons, orbit at distances of about ten fermis (roughly similar to the 8 fermi radius of the nucleus of uranium-238). These nuclei are not maximally dense. Halo nuclei form at the extreme edges of the chart of the nuclides—the neutron drip line and proton drip line—and are all unstable with short half-lives, measured in milliseconds; for example, lithium-11 has a half-life of less than 8.6 milliseconds. Halos in effect represent an excited state with nucleons in an outer quantum shell which has unfilled energy levels "below" it (both in terms of radius and energy). The halo may be made of either neutrons [NN, NNN] or protons [PP, PPP]. Examples: Nuclei which have a single neutron halo include 11Be and 19C. A two-neutron halo is exhibited by ^{6}He, ^{11}Li, ^{17}B, ^{19}B and ^{22}C. Two-neutron halo nuclei break into three fragments, never two, and are called *Borromean* because of this behaviour (referring to a system of three interlocked rings in which breaking any ring frees both of the others). ^{8}He and ^{14}Be both exhibit a four-neutron halo. Nuclei which have a proton halo include ^{8}B and ^{26}P. A two-proton halo is exhibited by ^{17}Ne and ^{27}S. Proton halos are expected to be more rare and unstable than the neutron examples, because of the repulsive electromagnetic forces of the excess proton(s).

Nuclear Models

There are many different historical models of the atomic nucleus, none of which to this day completely explains experimental data on nuclear structure. A useful review of 37 known models of the atomic nucleus is provided by Cook. In other words, packing protons and neutrons in the nucleus gives *approximately* the same total size result as packing hard spheres of a constant size (like marbles) into a tight spherical or semi-spherical bag (some stable nuclei are not quite spherical, but are known to be prolate).

Liquid Drop Models

Early models of the nucleus viewed the nucleus as a rotating liquid drop. In this model, the trade-off of long-range electromagnetic forces and relatively short-range nuclear forces, together cause behaviour which resembled surface tension forces in liquid drops of different sizes. This formula is successful at explaining many important phenomena of nuclei, such as their changing amounts of binding energy as their size and composition changes, but it does not explain the special stability which occurs when nuclei have special "magic numbers" of protons or neutrons.

Shell Models and Other Quantum Models

A number of models for the nucleus have also been proposed in which nucleons occupy orbitals, much like the atomic orbitals in atomic physics theory. These wave models imagine nucleons to be either sizeless point particles in potential wells, or else probability waves as in the "optical model", frictionlessly orbiting at high speed in potential wells. In these models, the nucleons may occupy orbitals in pairs, due to being fermions, but the exact nature and capacity of nuclear shells differs from those of electrons in atomic orbitals, primarily because the potential well in which the nucleons move (especially in larger nuclei) is quite different from the central electromagnetic potential well which binds electrons in atoms. Some resemblance to atomic orbital models may be seen in a small atomic nucleus like that of helium-4, in which the two protons and two neutrons separately occupy 1s orbitals analogous to the 1s orbital for the two electrons in the helium atom, and achieve unusual stability for the same reason. Nuclei with 5 nucleons are all extremely unstable and short-lived, yet, helium-3, with 3 nucleons, is very stable even with lack of a closed 1s orbital shell. Another nucleus with 3 nucleons, the triton hydrogen-3 is unstable and will decay into helium-3 when isolated. Weak nuclear stability with 2 nucleons {NP} in the 1s orbital is found in the deuteron hydrogen-2, with only one nucleon in each of the proton and neutron potential wells. While each nucleon is a fermion, the {NP} deuteron is a boson and thus does not follow Pauli Exclusion for close packing within shells. Lithium-6 with 6 nucleons is highly stable without a closed second 1p shell orbital. For light nuclei with total nucleon numbers 1 to 6 only those with 5 do not show some evidence of stability.

Observations of beta-stability of light nuclei outside closed shells indicate that nuclear stability is much more complex than simple closure of shell orbitals with magic numbers of protons and neutrons. For larger nuclei, the shells occupied by nucleons begin to differ significantly from electron shells, but nevertheless, present nuclear theory does predict the magic numbers of filled nuclear shells for both protons and neutrons.

The closure of the stable shells predicts unusually stable configurations, analogous to the noble group of nearly-inert gases in chemistry. An example is the stability of the closed shell of 50 protons, which allows tin to have 10 stable isotopes, more than any other element. Similarly, the distance from shell-closure explains the unusual instability of isotopes which have far from stable numbers of these particles, such as the radioactive elements 43 (technetium) and 61 (promethium), each of which is preceded and followed by 17 or more stable elements. There are however problems with the shell model when an attempt is made to account for nuclear properties well away from closed shells. This has led to complex *post hoc* distortions of the shape of the potential well to fit experimental data, but the question remains whether these mathematical manipulations actually correspond to the spatial deformations in real nuclei. Problems with the shell model have led some to propose realistic two-body and three-body nuclear force effects involving nucleon clusters and then build the nucleus on this basis. Two such cluster models are the Close-Packed Spheron Model of Linus Pauling and the 2D Ising Model of MacGregor.

Consistency between Models

As with the case of superfluid liquid helium, atomic nuclei are an example of a state in which both (1) "ordinary" particle physical rules for volume and (2) non-intuitive quantum mechanical rules for a wave-like nature apply. In superfluid helium, the helium atoms have volume, and essentially "touch" each other, yet at the same time exhibit strange bulk properties, consistent with a Bose-Einstein condensation. The latter reveals that they also have a wave-like nature and do not exhibit standard fluid properties, such as friction. For nuclei made of hadrons which are fermions, the same type of condensation does not occur, yet nevertheless, many nuclear properties can only be explained similarly by a combination of properties of particles with volume, in addition to the frictionless motion characteristic of the wave-like behaviour of objects trapped in Schrödinger quantum orbitals.

Background Radiation

Background radiation is radiation that is constantly present in the environment and is emitted from a variety of natural and artificial sources. Primary contributions come from:

- Sources in the earth. These include sources in food and water, which are incorporated in the body, and in building materials and other products that incorporate those radioactive sources;
- Sources from space, in the form of cosmic rays;
- Sources in the atmosphere. One significant contribution comes from the radon gas that is released from the Earth's crust and subsequently decays into radioactive atoms that become attached to airborne dust and particulates. Another contribution arises from the radioactive atoms produced in the bombardment of atoms in the upper atmosphere by high-energy cosmic rays.

Naturally occurring sources are responsible for the vast majority of radiation exposure. However, not including direct exposure from radiological imaging or therapy, about 3% of background radiation comes from man-made sources such as:

- Self-luminous dials and signs
- Global radioactive contamination due to historical nuclear weapons testing
- Nuclear power station or nuclear fuel reprocessing accidents

- Normal operation of facilities used for nuclear power and scientific research
- Emissions from burning fossil fuels, such as coal fired power plants
- Emissions from nuclear medicine facilities and patients

Natural Background Radiation

Natural background radiation comes from two primary sources: cosmic radiation and terrestrial sources. The worldwide average background dose for a human being is about 2.4 millisievert (mSv) per year. This exposure is mostly from cosmic radiation and natural radionuclides in the environment (including those within the body). This is far greater than human-caused background radiation exposure, which in the year 2000 amounted to an average of about 5 ìSv per year from historical nuclear weapons testing, nuclear power accidents and nuclear industry operation combined, and is greater than the average exposure from medical tests, which ranges from 0.04 to 1 mSv per year. Older coal-fired power plants without effective fly ash capture are one of the largest sources of human-caused background radiation exposure. The level of natural background radiation varies depending on location, and in some areas the level is significantly higher than average. Such areas include Ramsar in Iran, Guarapari in Brazil, Kerala in India, the northern Flinders Ranges in Australia and Yangjiang in China. In Ramsar a peak yearly dose of 260 mSv has been reported (compared with 0.06 mSv of a chest radiograph or up to 20 mSv of a CT scan).

The highest levels of natural background radiation recorded in the world is from areas around Ramsar, particularly at Talesh-Mahalleh which is a very high background radiation area (VHBRA) having an effective dose equivalent several times in excess of ICRP-recommended radiation dose limits for radiation workers and up to 200 times greater than normal background levels. Most of the radiation in the area is due to dissolved radium-226 in water of hot springs along with smaller amounts of uranium and thorium due to travertine deposits. There are more than nine hot springs in the area with different concentrations of radioisotopes, and these are used as spas by locals and tourists. This high level of radiation does not seem to have caused ill effects on the residents of the area and even possibly has made them slightly more radioresistant, which is puzzling and has been called "radiation paradox". It has also been reported that residents have healthier and longer lives. On the basis of this and other evidences including the fact that life had originated in a much more irradiated environment, some scientists have questioned the validity of linear no-threshold model, on which all radiation regulations currently depend. Others point out that some level of radiation might actually be good for health and have a positive effect on population based on the controversial radiation hormesis model, by jump starting DNA repair mechanisms inside the cell. Background radiation doses in the immediate vicinities of particles of high atomic number materials, within the human body, have a small enhancement due to the photoelectric effect.

Cosmic Radiation

The Earth, and all living things on it, are constantly bombarded by radiation from outer space. This radiation primarily consists of positively charged ions from protons to iron and larger nuclei derived sources outside our solar system. This radiation interacts with atoms in the atmosphere to create secondary radiation, including X-rays, muons, protons, alpha particles, pions, electrons, and neutrons. The immediate dose from cosmic radiation is largely from muons, neutrons, and electrons, and this dose varies in different parts of the world based largely on the geomagnetic field and altitude. This radiation is much more intense in the upper troposphere, around 10 km altitude, and is thus of particular concern for airline crews and frequent passengers, who spend many hours per year in this environment. During their flights airline crews typically get an extra dose on the order of 2.2mSv (220 mrem) per year. Similarly, cosmic rays cause higher background exposure in astronauts than in humans on the surface of Earth. Astronauts in low orbits, such as in the International Space Station or the Space Shuttle, are partially shielded by the magnetic field of the Earth, but also suffer from the Van Allen radiation belt which accumulates cosmic rays and results from the earths magnetic field. Outside low Earth orbit, as experienced by the Apollo astronauts who traveled to the Moon, this background radiation is much more intense, and represents a considerable obstacle to potential future long term human exploration of the moon or Mars.

Cosmic rays also cause elemental transmutation in the atmosphere, in which secondary radiation generated by the cosmic rays combines with atomic nuclei in the atmosphere to generate different nuclides. Many so-called cosmogenic nuclides can be produced, but probably the most notable is carbon-14, which is produced by interactions with nitrogen atoms. These cosmogenic nuclides eventually reach the Earth's surface and can be incorporated into living organisms. The production of these nuclides varies slightly with short-term variations in solar cosmic ray flux, but is considered practically constant over long scales of thousands to millions of years. The constant production, incorporation into organisms and relatively short half-life of carbon-14 are the principles used in radiocarbon dating of ancient biological materials such as wooden artifacts or human remains.

Terrestrial Sources

Radioactive material is found throughout nature. It occurs naturally in the soil, rocks, water, air, and vegetation. The major radionuclides of concern for terrestrial radiation are common elements with low-abundance radioactive isotopes, like potassium and carbon, or the long-lived elements uranium and thorium and their decay products, some of which, like radium and radon are intensely radioactive but occur in low concentration. Most of these sources have been decreasing, due to radioactive decay since the formation of the Earth, because there is no significant amount currently transported to the Earth. Thus, the present activity on earth from uranium-238 is only half as much as it originally was because of its 4.5 billion year half-life, and potassium-40 (half life 1.25 billion years) is only at about 8% of original activity. The effects on humas of the actual diminishment (due to decay) of these isotopes is minimal however. This is because humans evolved too recently for the difference in activity over a fraction of a half-life to be significant. Put another way, human history is so short in comparison to a half life of a billion years, that the activity of these long-lived isotopes has been effectively constant throughout our time on this planet. In addition, many shorter half-life and thus more intensely radioactive isotopes have not decayed out of the terrestrial environment, however, because of natural on-going production of them. Examples of these are carbon-14 (cosmogenic), radium-226 (decay product of uranium-238) and radon-222 (a decay product of radium-226).

Radiation Inside the Human Body

Some of the essential elements that make up the human body, mainly potassium and carbon, have radioactive isotopes that add significantly to our background radiation dose. An average human contains about 30 milligrams of potassium-40 (^{40}K) and about 10 nanograms (10^{-8} g) of carbon-14 (^{14}C). Excluding internal contamination by external radioactive material, the largest component of internal radiation exposure from biologically functional components of the human body is from potassium-40.

The decay of about 4,000 nuclei of ^{40}K per second makes potassium the largest source of radiation in terms of number of decaying atoms. The energy of beta particles produced by ^{40}K is also about 10 times more powerful than the beta particles from ^{14}C decay. There are about 1,200 beta particles per second produced by the decay of ^{14}C. However, a ^{14}C atom is in the genetic information of about half the cells, while potassium is not a component of DNA. The decay of a ^{14}C atom inside DNA in one person happens about 50 times per second, changing a carbon atom to one of nitrogen.

Radon

Radon is a terrestrial source of ionizing radiation that is of particular concern because, although on average it is very rare, this intensely radioactive element can be found in high concentrations in many areas of the world, where it represents a significant health hazard. Radon is a decay product of uranium, which is relatively common in the Earth's crust, but generally concentrated in ore-bearing rocks scattered around the world.

Radon seeps out of these ores into the atmosphere or into ground water, and in these localities it can accumulate within dwellings and expose humans to high concentrations. The widespread construction of well insulated and sealed homes in the northern industrialized world has led to radon becoming the primary source of background radiation in some localities in northern North America and Europe. Some of these areas, including Cornwall and Aberdeenshire in the United Kingdom have high enough natural radiation levels that nuclear licensed sites cannot be built there — the sites would already exceed legal radiation limits before they opened, and the natural topsoil and rock would all have to be disposed of as low-level nuclear waste. Radiation exposure from radon is indirect. Radon has a short half-life (4 days) and decays into other solid particulate radium-series radioactive nuclides. These radioactive particles are inhaled and remain lodged in the lungs, causing continued exposure. People in affected localities can receive up to 10 mSv per year background radiation. Radon is thus the second leading cause of lung cancer after smoking, and accounts for 15,000 to 22,000 cancer deaths per year in the US alone.

Human-caused Background Radiation

Frequent above-ground nuclear explosions between the 1940s and 1960s scattered a substantial amount of radioactive contamination. Some of this contamination is local, rendering the immediate surroundings highly radioactive, while some of it is carried longer distances as nuclear fallout; some of this material is dispersed worldwide. The increase in background radiation due to these tests peaked in 1963 at about 0.15 mSv per year worldwide, or about 7% of average background dose from all sources.

The Limited Test Ban Treaty of 1963 prohibited above-ground tests, thus by the year 2000 the worldwide dose from these historical tests has decreased to only 0.005 mSv per year. Older coal-fired power plants without effective fly ash capture are one of the largest sources of human-caused background radiation exposure. When coal is burned, uranium, thorium and all the uranium daughters accumulated by disintegration — radium, radon, polonium — are released. According to a 1978 article in Science magazine, "coal-fired power plants throughout the world are the major sources of radioactive materials released to the environment". Radioactive materials previously buried underground in coal deposits are released as fly ash or, if fly ash is captured, may be incorporated into concrete manufactured with fly ash. Radioactive materials are also released in gaseous emissions.

The United Nations Scientific Committee on the Effects of Atomic Radiation estimates that per gigawatt-year ($GW_e a$) of electrical energy produced by coal, using the current mix of technology throughout the world, the population impact is approximately 0.8 lethal cancers per plant-year distributed over the affected population.

With 400 GW of coal-fired power plants in the world, this amounts to some 320 deaths per year. Nuclear reactors may also release a certain amount of radioactive contamination. Under normal circumstances, a modern nuclear reactor releases minuscule amounts of radioactive contamination. Major accidents, which have been rare, have also released some radioactive contamination into the environment; this was the case with the Windscale fire (Sellafield accident), the Chernobyl accident, and the Fukushima I nuclear accidents. The amount of radioactive contamination released by human activity is rather small and in most cases negligible in comparison to natural background radiation.

Artificial Radiation Sources

The radiation from natural and artificial radiation sources are identical in their nature and their effects. These materials are distributed in the environment, and in our bodies, according to the chemical properties of the elements. The Nuclear Regulatory Commission, the United States Environmental Protection Agency, and other U.S. and international agencies, require that licensees limit radiation exposure to individual members of the public to 1 mSv (100 mrem) per year, and limit occupational radiation exposure to adults working with radioactive material to 50 mSv (5 rem) per year, and 100 mSv (10 rem) in 5 years. The exposure for an average person is about 3.6 mSv/year, 80 percent of which comes from natural sources of radiation. The remaining 20 percent results from exposure to artificial radiation sources, such as medical x-rays, industrial sources like smoke detectors and a small fraction from nuclear weapons tests. For average persons who have had no medical x-rays, only 3% of their annual radiation dose comes from artificial sources. A standard medical x-ray's strength is about 2 mrem or 0.02 mSv but can be over ten times that, depending on the equipment used. A dental x-ray optimally has a dose as low as 0.0033 mSv but poor machines and technique can give doses as high as 0.11 mSv. The average American and European receives about 0.5 mSv of diagnostic medical dose per year; countries with lower levels of health care receive about one fifth of this dose. Radiation treatment for various diseases also accounts for some dose, both in individuals and in those around them.

Other Usage

In other contexts, background radiation may simply be any radiation that is pervasive, whether ionizing or not. A particular example of this is the cosmic microwave background radiation, a nearly uniform glow that fills the sky in the microwave part of the spectrum; stars, galaxies and other objects of interest in radio astronomy stand out against this background. In a laboratory, background radiation refers to the measured value from any sources that affect an instrument when a radiation source sample is not being measured. This background rate, which must be established as a stable value by multiple measurements, usually before and after sample measurement, is subtracted from the rate measured when the sample is being measured. Background radiation for occupational doses measured for workers is all radiation dose that is not measured by radiation dose measurement instruments in potential occupational exposure conditions. This includes both "natural background radiation" and any medical radiation doses. This value is not typically measured or known from surveys, such that variations in the total dose to individual workers is not known. This can be a significant confounding factor in assessing radiation exposure effects in a population of workers who may have significantly different natural background and medical radiation doses. This is most significant when the occupational doses are very low.

Magnetic Resonance Imaging

Magnetic resonance imaging (MRI), nuclear magnetic resonance imaging (NMRI), or magnetic resonance tomography (MRT) is a medical imaging technique used in radiology to visualize detailed internal structures. MRI makes use of the property of Nuclear magnetic resonance (NMR) to image nuclei of atoms inside the body. An MRI machine uses a powerful magnetic field to align the magnetization of some atoms in the body, and radio frequency fields to systematically alter the alignment of this magnetization. This causes the nuclei to produce a rotating magnetic field detectable by the scanner—and this information is recorded to construct an image of the scanned area of the body. Strong magnetic field gradients cause nuclei at different locations to rotate at different speeds. 3-D spatial information can be obtained by providing gradients in each direction. MRI provides good contrast between the different soft tissues of the body, which make it especially useful in imaging the brain, muscles, the heart, and cancers compared with other medical imaging techniques such as computed tomography (CT) or X-rays. Unlike CT scans or traditional X-rays, MRI uses no ionizing radiation.

How MRI Works

The body is largely composed of water molecules. Each water molecule has two hydrogen nuclei or protons. When a person goes inside the powerful magnetic field of the scanner, the magnetic moments of some of these protons change and align with the direction of the field. In an MRI machine a radio frequency transmitter is briefly turned on, producing an electromagnetic field. The photons of this field have just the right energy, known as the resonance frequency, to flip the spin of the aligned protons in the body. As the intensity and duration of application of the field increase, more aligned spins are affected. After the field is turned off, the protons decay to the original spin-down state and the difference in energy between the two states is released as a photon. It is these photons that produce the electromagnetic signal that the scanner detects. The frequency at which the protons resonate depends on the strength of the magnetic field. As a result of conservation of energy, this resonation frequency also dictates the frequency of the released photons. The photons released when the field is removed have an energy — and therefore a frequency — due to the amount of energy the protons absorbed while the field was active. It is this relationship between field-strength and frequency that allows the use of nuclear magnetic resonance for imaging. Additional magnetic fields are applied during the scan to make the magnetic field strength depend on the position within the patient, in turn making the frequency of the released photons dependent on position in a predictable manner. Position information can then be recovered from the resulting signal by the use of a Fourier transform.

These fields are created by passing electric currents through specially-wound solenoids, known as gradient coils. Since these coils are within the bore of the scanner, there are large forces between them and the main field coils, producing most of the noise that is heard during operation. Without efforts to damp this noise, it can approach 130 decibels (dB) with strong fields . An image can be constructed because the protons in different tissues return to their equilibrium state at different rates, which is a difference that can be detected. Five different tissue variables — spin density, T_1 and T_2 relaxation times and flow and spectral shifts can be used to construct images. By changing the parameters on the scanner, this effect is used to create contrast between different types of body tissue or between other properties, as in fMRI and diffusion MRI. Contrast agents may be injected intravenously to enhance the appearance of blood vessels, tumors or inflammation. Contrast agents may also be directly injected into a joint in the case of arthrograms, MRI images of joints. Unlike CT, MRI uses no ionizing radiation and is generally a very safe procedure. Nonetheless the strong magnetic fields and radio pulses can affect metal implants, including cochlear implants and cardiac pacemakers. In the case of cochlear implants, the US FDA has approved some implants for MRI compatibility. In the case of cardiac pacemakers, the results can sometimes be lethal, so patients with such implants are generally not eligible for MRI. MRI is used to image every part of the body, and is particularly useful for tissues with many hydrogen nuclei and little density contrast, such as the brain, muscle, connective tissue and most tumors.

History

Magnetic resonance imaging is a relatively new technology. The first MR image was published in 1973 and the first cross-sectional image of a living mouse was published in January 1974. The first studies performed on humans were published in 1977. By comparison, the first human X-ray image was taken in 1895.

Applications

In clinical practice, MRI is used to distinguish pathologic tissue (such as a brain tumor) from normal tissue. One advantage of an MRI scan is that it is harmless to the patient. It uses strong magnetic fields and non-ionizing radiation in the radio frequency range, unlike CT scans and traditional X-rays, which both use ionizing radiation. While CT provides good spatial resolution (the ability to distinguish two separate structures an arbitrarily small distance from each other), MRI provides comparable resolution with far better contrast resolution (the ability to distinguish the differences between two arbitrarily similar but not identical tissues).

The basis of this ability is the complex library of *pulse sequences* that the modern medical MRI scanner includes, each of which is optimized to provide *image contrast* based on the chemical sensitivity of MRI. For example, with particular values of the *echo time* (T_E) and the *repetition time* (T_R), which are basic parameters of image acquisition, a sequence takes on the property of T_2-weighting. On a T_2-weighted scan, water- and fluid-containing tissues are bright (most modern T_2 sequences are actually *fast* T_2 sequences) and fat-containing tissues are dark.

The reverse is true for T_1-weighted images. Damaged tissue tends to develop edema, which makes a T_2-weighted sequence sensitive for pathology, and generally able to distinguish pathologic tissue from normal tissue. With the addition of an additional radio frequency pulse and additional manipulation of the magnetic gradients, a T_2-weighted sequence can be converted to a FLAIR sequence, in which free water is now dark, but edematous tissues remain bright. This sequence in particular is currently the most sensitive way to evaluate the brain for demyelinating diseases, such as multiple sclerosis. The typical MRI examination consists of 5–20 sequences, each of which are chosen to provide a particular type of information about the subject tissues. This information is then synthesized by the interpreting physician.

Basic MRI Scans

T_1-Weighted MRI: T_1-weighted scans are a standard basic scan, in particular differentiating fat from water - with water darker and fat brighter use a gradient echo (GRE) sequence, with short T_E and short T_R. This is one of the basic types of MR contrast and is a commonly run clinical scan. The T_1 weighting can be increased (improving contrast) with the use of an inversion pulse as in an MP-RAGE sequence. Due to the short repetition time (T_R) this scan can be run very fast allowing the collection of high resolution 3D datasets. A T_1 reducing gadolinium contrast agent is also commonly used, with a T_1 scan being collected before and after administration of contrast agent to compare the difference. In the brain T_1-weighted scans provide good gray matter/white matter contrast; in other words, T_1-weighted images highlight fat deposition.

Spin Density Weighted MRI

Spin density, also called proton density, weighted scans try to have no contrast from either T_2 or T_1 decay, the only signal change coming from differences in the amount of available spins (hydrogen nuclei in water). It uses a spin echo or sometimes a gradient echo sequence, with short T_E and long T_R.

Specialized MRI Scans

Diffusion MRI: Diffusion MRI measures the diffusion of water molecules in biological tissues. In an isotropic medium (inside a glass of water for example) water molecules naturally move randomly according to turbulence and Brownian motion. In biological tissues however, where the Reynold's number is low enough for flows to be laminar, the diffusion may be anisotropic. For example a molecule inside the axon of a neuron has a low probability of crossing the myelin membrane. Therefore the molecule moves principally along the axis of the neural fiber. If we know that molecules in a particular voxel diffuse principally in one direction we can make the assumption that the majority of the fibers in this area are going parallel to that direction. The recent development of diffusion tensor imaging (DTI) enables diffusion to be measured in multiple directions and the fractional anisotropy in each direction to be calculated for each voxel. This enables researchers to make brain maps of fiber directions to examine the connectivity of different regions in the brain (using tractography) or to examine areas of neural degeneration and demyelination in diseases like Multiple Sclerosis. Another application of diffusion MRI is diffusion-weighted imaging (DWI). Following an ischemic stroke, DWI is highly sensitive to the changes occurring in the lesion. It is speculated that increases in restriction (barriers) to water diffusion, as a result of cytotoxic edema (cellular swelling), is responsible for the increase in signal on a DWI scan. The DWI enhancement appears within 5–10 minutes of the onset of stroke symptoms (as compared with computed tomography, which often does not detect changes of acute infarct for up to 4–6 hours) and remains for up to two weeks. Coupled with imaging of cerebral perfusion, researchers can highlight regions of "perfusion/diffusion mismatch" that may indicate regions capable of salvage by reperfusion therapy. Like many other specialized applications, this technique is usually coupled with a fast image acquisition sequence, such as echo planar imaging sequence.

Magnetization Transfer MRI: Magnetization transfer (MT) refers to the transfer of longitudinal magnetization from free water protons to hydration water protons in NMR and MRI. In magnetic resonance imaging of molecular solutions, such as protein solutions, two types of water molecules, free (bulk) and hydration (bound), are found. Free water protons have faster average rotational frequency and hence less fixed water molecules that may cause local field inhomogeneity. Because of this uniformity, most free water protons have resonance frequency lying narrowly around the normal proton resonance frequency of 63 MHz (at 1.5 teslas). This also results in slower transverse magnetization dephasing and hence longer T_2. Conversely, hydration water molecules are slowed down by interaction with solute molecules and hence create field inhomogeneities that lead to wider resonance frequency spectrum. In free liquids, protons, which may be viewed classically as small magnetic dipoles, exhibit translational and rotational motions. These moving dipoles disturb the surrounding magnetic field however on long enough time-scales (which may be nanoseconds) the average field caused by the motion of protons is zero. This is known as "motional averaging" or narrowing and is characteristic of protons moving freely in liquid. On the other hand, protons bound to macromolecules, such as proteins, tend to have a fixed orientation and so the average magnetic field in close proximity to such structures does not average to zero. The result is a spatial pattern in the magnetic field that gives rise to a residual dipolar coupling (range of precession frequencies) for the protons experiencing the magnetic field.

The wide frequency distribution appears as a broad spectrum that may be several kHz wide. The net signal from these protons disappears very quickly, in inverse proportion to the width, due to the loss of coherence of the spins, i.e. T2 relaxation. Due to exchange mechanisms, such as spin transfer or proton chemical exchange, the (incoherent) spins bound to the macromolecules continually switch places with (coherent) spins in the bulk media and establish a dynamic equilibrium. Magnetization transfer: Although there is no measurable signal from the bound spins, or the bound spins that exchange into the bulk media, their longitudinal magnetization is preserved and may recover only via the relatively slow process of T1 relaxation. If the longitudinal magnetization of just the bound spins can be altered, then the effect can be measured in the spins of the bulk media due to the exchange processes. The magnetization transfer sequence applies RF saturation at a frequency that is far off resonance for the narrow line of bulk water but still on resonance for the bound protons with a spectral linewidth of kHz. This causes saturation of the bound spins which exchange into the bulk water, resulting in a loss of longitudinal magnetization and hence signal decrease in the bulk water. This provides an indirect measure of macromolecular content in tissue. Implementation of magnetization transfer involves choosing suitable frequency offsets and pulse shapes to saturate the bound spins sufficiently strongly, within the safety limits of specific absorption rate for RF irradiation.

T1ρ (T1rho): Molecules have a kinetic energy that is a function of the temperature and is expressed as translational and rotational motions, and by collisions between molecules. The moving dipoles disturb the magnetic field but are often extremely rapid so that the average effect over a long time-scale may be zero. However, depending on the time-scale, the interactions between the dipoles do not always average away. At the slowest extreme the interaction time is effectively infinite and occurs where there are large, stationary field disturbances (e.g. a metallic implant). In this case the loss of coherence is described as a "static dephasing". T2* is a measure of the loss of coherence in an ensemble of spins that include all interactions (including static dephasing). T2 is a measure of the loss of coherence that excludes static dephasing, using an RF pulse to reverse the slowest types of dipolar interaction. There is in fact a continuum of interaction time-scales in a given biological sample and the properties of the refocusing RF pulse can be tuned to refocus more than just static dephasing. In general, the rate of decay of an ensemble of spins is a function of the interaction times and also the power of the RF pulse. This type of decay, occurring under the influence of RF, is known as T1ρ. It is similar to T2 decay but with some slower dipolar interactions refocused as well as the static interactions.

Fluid Attenuated Inversion Recovery (FLAIR): Fluid Attenuated Inversion Recovery (FLAIR) is an inversion-recovery pulse sequence used to null signal from fluids. For example, it can be used in brain imaging to suppress cerebrospinal fluid (CSF) so as to bring out the periventricular hyperintense lesions, such as multiple sclerosis (MS) plaques. By carefully choosing the inversion time TI (the time between the inversion and excitation pulses), the signal from any particular tissue can be suppressed.

Magnetic Resonance Angiography

Magnetic resonance angiography (MRA) generates pictures of the arteries to evaluate them for stenosis (abnormal narrowing) or aneurysms (vessel wall dilatations, at risk of rupture). MRA is often used to evaluate the arteries of the neck and brain, the thoracic and abdominal aorta, the renal arteries, and the legs (called a "run-off"). A variety of techniques can be used to generate the pictures, such as administration of a paramagnetic contrast agent (gadolinium) or using a technique known as "flow-related enhancement" (e.g. 2D and 3D time-of-flight sequences), where most of the signal on an image is due to blood that recently moved into that plane, see also FLASH MRI. Techniques involving phase accumulation (known as phase contrast angiography) can also be used to generate flow velocity maps easily and accurately. Manetic resonance venography (MRV) is a similar procedure that is used to image veins. In this method, the tissue is now excited inferiorly, while signal is gathered in the plane immediately superior to the excitation plane—thus imaging the venous blood that recently moved from the excited plane.

Magnetic Resonance Gated Intracranial CSF Dynamics (MR-GILD): Magnetic resonance gated intracranial cerebrospinal fluid (CSF) or liquor dynamics (MR-GILD) technique is an MR sequence based on bipolar gradient pulse used to demonstrate CSF pulsatile flow in ventricles, cisterns, aqueduct of Sylvius and entire intracranial CSF pathway. It is a method for analyzing CSF circulatory system dynamics in patients with CSF obstructive lesions such as normal pressure hydrocephalus. It also allows visualization of both arterial and venous pulsatile blood flow in vessels without use of contrast agents.

Magnetic Resonance Spectroscopy: Magnetic resonance spectroscopy (MRS) is used to measure the levels of different metabolites in body tissues. The MR signal produces a spectrum of resonances that correspond to different molecular arrangements of the isotope being "excited". This signature is used to diagnose certain metabolic disorders, especially those affecting the brain, and to provide information on tumor metabolism. Magnetic resonance spectroscopic imaging (MRSI) combines both spectroscopic and imaging methods to produce spatially localized spectra from within the sample or patient. The spatial resolution is much lower (limited by the available SNR), but the spectra in each voxel contains information about many metabolites. Because the available signal is used to encode spatial and spectral information, MRSI requires high SNR achievable only at higher field strengths (3 T and above).

Functional MRI: Functional MRI (fMRI) measures signal changes in the brain that are due to changing neural activity. The brain is scanned at low resolution but at a rapid rate (typically once every 2–3 seconds). Increases in neural activity cause changes in the MR signal via T_2 changes; this mechanism is referred to as the BOLD (blood-oxygen-level dependent) effect. Increased neural activity causes an increased demand for oxygen, and the vascular system actually overcompensates for this, increasing the amount of oxygenated hemoglobin relative to deoxygenated hemoglobin. Because deoxygenated hemoglobin attenuates the MR signal, the vascular response leads to a signal increase that is related to the neural activity. The precise nature of the relationship between neural activity and the BOLD signal is a subject of current research. The BOLD effect also allows for the generation of high resolution 3D maps of the venous vasculature within neural tissue. While BOLD signal is the most common method employed for neuroscience studies in human subjects, the flexible nature of MR imaging provides means to sensitize the signal to other aspects of the blood supply. Alternative techniques employ arterial spin labeling (ASL) or weight the MRI signal by cerebral blood flow (CBF) and cerebral blood volume (CBV). The CBV method requires injection of a class of MRI contrast agents that are now in human clinical trials. Because this method has been shown to be far more sensitive than the BOLD technique in preclinical studies, it may potentially expand the role of fMRI in clinical applications. The CBF method provides more quantitative information than the BOLD signal, albeit at a significant loss of detection sensitivity.

Real-time MRI: Real-time MRI refers to the continuous monitoring ("filming") of moving objects in real time. While many different strategies have been developed over the past two decades, a recent development reported a real-time MRI technique based on radial FLASH that yields a temporal resolution of 20 to 30 milliseconds for images with an in-plane resolution of 1.5 to 2.0 mm. The new method promises to add important information about diseases of the joints and the heart. In many cases MRI examinations may become easier and more comfortable for patients.

Interventional MRI: The lack of harmful effects on the patient and the operator make MRI well-suited for "interventional radiology", where the images produced by a MRI scanner are used to guide minimally invasive procedures. Of course, such procedures must be done without *any* ferromagnetic instruments. A specialized growing subset of interventional MRI is that of intraoperative MRI in which the MRI is used in the surgical process. Some specialized MRI systems have been developed that allow imaging concurrent with the surgical procedure. More typical, however, is that the surgical procedure is temporarily interrupted so that MR images can be acquired to verify the success of the procedure or guide subsequent surgical work.

Radiation Therapy Simulation

Because of MRI's superior imaging of soft tissues, it is now being utilized to specifically locate tumors within the body in preparation for radiation therapy treatments. For therapy simulation, a patient is placed in specific, reproducible, body position and scanned. The MRI system then computes the precise location, shape and orientation of the tumor mass, correcting for any spatial distortion inherent in the system. The patient is then marked or tattooed with points that, when combined with the specific body position, permits precise triangulation for radiation therapy.

Current Density Imaging: Current density imaging (CDI) endeavors to use the phase information from images to reconstruct current densities within a subject. Current density imaging works because electrical currents generate magnetic fields, which in turn affect the phase of the magnetic dipoles during an imaging sequence.

Magnetic Resonance Guided Focused Ultrasound: In MRgFUS therapy, ultrasound beams are focused on a tissue—guided and controlled using MR thermal imaging—and due to the significant energy deposition at the focus, temperature within the tissue rises to more than 65 °C (150 °F), completely destroying it. This technology can achieve precise ablation of diseased tissue. MR imaging provides a three-dimensional view of the target tissue, allowing for precise focusing of ultrasound energy. The MR imaging provides quantitative, real-time, thermal images of the treated area. This allows the physician to ensure that the temperature generated during each cycle of ultrasound energy is sufficient to cause thermal ablation within the desired tissue and if not, to adapt the parameters to ensure effective treatment. This allows the physician to ensure that the temperature generated during each cycle of ultrasoound energy is sufficient to cause thermal ablation within the desired tissue and if not, to adapt the parameters to ensure effective treatment.

Multinuclear Imaging: Hydrogen is the most frequently imaged nucleus in MRI because it is present in biological tissues in great abundance, and because its high gyromagnetic ratio gives a strong signal. However, any nucleus with a net nuclear spin could potentially be imaged with MRI. Such nuclei include helium-3, carbon-13, fluorine-19, oxygen-17, sodium-23, phosphorus-31 and xenon-129. ^{23}Na and ^{31}P are naturally abundant in the body, so can be imaged directly. Gaseous isotopes such as ^{3}He or ^{129}Xe must be hyperpolarized and then inhaled as their nuclear density is too low to yield a useful signal under normal conditions. ^{17}O and ^{19}F can be administered in sufficient quantities in liquid form (e.g. ^{17}O-water) that hyperpolarization is not a necessity. Multinuclear imaging is primarily a research technique at present. However, potential applications include functional imaging and imaging of organs poorly seen on ^{1}H MRI (e.g. lungs and bones) or as alternative contrast agents. Inhaled hyperpolarized ^{3}He can be used to image the distribution of air spaces within the lungs. Injectable solutions containing ^{13}C or stabilized bubbles of hyperpolarized ^{129}Xe have been studied as contrast agents for angiography and perfusion imaging. ^{31}P can potentially provide information on bone density and structure, as well as functional imaging of the brain.

Susceptibility Weighted Imaging (SWI): Susceptibility weighted imaging (SWI), is a new type of contrast in MRI different from spin density, T_1, or T_2 imaging. This method exploits the susceptibility differences between tissues and uses a fully velocity compensated, three dimensional, RF spoiled, high-resolution, 3D gradient echo scan. This special data acquisition and image processing produces an enhanced contrast magnitude image very sensitive to venous blood, hemorrhage and iron storage. It is used to enhance the detection and diagnosis of tumors, vascular and neurovascular diseases (stroke and hemorrhage, multiple sclerosis, Alzheimer's), and also detects traumatic brain injuries that may not be diagnosed using other methods

Other Specialized MRI Techniques: Field of research and new methods and variants are often published when they are able to get better results in specific fields. Examples of these recent improvements are T_2-weighted turbo spin-echo (T_2 TSE MRI), double inversion recovery MRI (DIR-MRI) or phase-sensitive inversion recovery MRI (PSIR-MRI), all of them able to improve imaging of the brain lesions. Another example is MP-RAGE (magnetization-prepared rapid acquisition with gradient echo), which improves images of multiple sclerosis cortical lesions.

Portable Instruments

Portable magnetic resonance instruments are available for use in education and field research. Using the principles of Earth's field NMR, they have no powerful polarizing magnet, so that such instruments can be small and inexpensive. Some can be used for both EFNMR spectroscopy and MRI imaging. The low strength of the Earth's field results in poor signal to noise ratios, requiring long scan times to capture spectroscopic data or build up MRI images. Research with atomic magnetometers have discussed the possibility for cheap and portable MRI instruments without the large magnet.

MRI versus CT

A computed tomography (CT) scanner uses X-rays, a type of ionizing radiation, to acquire its images, making it a good tool for examining tissue composed of elements of a higher atomic number than the tissue surrounding them, such as bone and calcifications (calcium based) within the body (carbon based flesh), or of structures (vessels, bowel). MRI, on the other hand, uses non-ionizing radio frequency (RF) signals to acquire its images and is best suited for non-calcified tissue, though MR images can also be acquired from bones and teeth as well as fossils. CT may be enhanced by use of contrast agents containing elements of a higher atomic number than the surrounding flesh such as iodine or barium. Contrast agents for MRI have paramagnetic properties, e.g., gadolinium and manganese. Both CT and MRI scanners are able to generate multiple two-dimensional cross-sections (slices) of tissue and three-dimensional reconstructions. Unlike CT, which uses only X-ray attenuation to generate image contrast, MRI has a long list of properties that may be used to generate image contrast. By variation of scanning parameters, tissue contrast can be altered and enhanced in various ways to detect different features. MRI can generate cross-sectional images in any plane (including oblique planes). In the past, CT was limited to acquiring images in the axial (or near axial) plane. The scans used to be called Computed *Axial* Tomography scans (CAT scans). However, the development of multi-detector CT scanners with near-isotropic resolution, allows the CT scanner to produce data that can be retrospectively reconstructed in any plane with minimal loss of image quality. For purposes of tumor detection and identification in the brain, MRI is generally superior. However, in the case of solid tumors of the abdomen and chest, CT is often preferred due to less motion artifact. Furthermore, CT usually is more widely available, faster, less expensive, and may be less likely to require the person to be sedated or anesthetized. MRI is also best suited for cases when a patient is to undergo the exam several times successively in the short term, because, unlike CT, it does not expose the patient to the hazards of ionizing radiation.

Economics of MRI

MRI equipment is expensive. 1.5 tesla scanners often cost between $1 million and $1.5 million USD. 3.0 tesla scanners often cost between $2 million and $2.3 million USD. Construction of MRI suites can cost up to $500,000 USD, or more, depending on project scope. MRI scanners have been significant sources of revenue for healthcare providers in the US. This is because of favourable reimbursement rates from insurers and federal government programmes. Insurance reimbursement is provided in two components, an equipment charge for the actual performance of the MRI scan and professional charge for the radiologist's review of the images and/or data. In the US Northeast, an equipment charge might be $3,500 and a professional charge might be $350 although the actual fees received by the equipment owner and interpreting physician are often significantly less and depend on the rates negotiated with insurance companies or determined by governmental action as in the Medicare Fee Schedule. For example, an orthopedic surgery group in Illinois billed a charge of $1,116 for a knee MRI in 2007 but the Medicare reimbursement in 2007 was only $470.91. Many insurance companies require preapproval of an MRI procedure as a condition for coverage. In the US, the Deficit Reduction Act of 2007 significantly reduced reimbursement rates paid by federal insurance programmes for the equipment component of many scans, shifting the economic landscape. Many private insurers have followed suit.

Safety

A number of features of MRI scanning can give rise to risks. These include:

- Powerful magnetic fields
- Cryogenic liquids
- Noise
- Claustrophobia

In addition, in cases where MRI contrast agents are used, these also typically have associated risks.

Magnetic Field

Most forms of medical or biostimulation implants are generally considered contraindications for MRI scanning. These include pacemakers, vagus nerve stimulators, implantable cardioverter-defibrillators, loop recorders, insulin pumps, cochlear implants, deep brain stimulators and capsules retained from capsule endoscopy. Patients are therefore always asked for complete information about all implants before entering the room for an MRI scan. Several deaths have been reported in patients with pacemakers who have undergone MRI scanning without appropriate precautions. To reduce such risks, implants are increasingly being developed to make them able to be safely scanned, and specialized protocols have been developed to permit the safe scanning of selected implants and pacing devices. Ferromagnetic foreign bodies such as shell fragments, or metallic implants such as surgical prostheses and aneurysm clips are also potential risks. Interaction of the magnetic and radio frequency fields with such objects can lead to trauma due to movement of the object in the magnetic field or thermal injury from radio-frequency induction heating of the object. Titanium and its alloys are safe from movement from the magnetic field. In the United States a classification system for implants and ancillary clinical devices has been developed by ASTM International and is now the standard supported by the US Food and Drug Administration:

- *MR-Safe* — The device or implant is completely non-magnetic, non-electrically conductive, and non-RF reactive, eliminating all of the primary potential threats during an MRI procedure.
- *MR-Conditional* — A device or implant that may contain magnetic, electrically conductive or RF-reactive components

that is safe for operations in proximity to the MRI, provided the conditions for safe operation are defined and observed (such as 'tested safe to 1.5 teslas' or 'safe in magnetic fields below 500 gauss in strength').

- *MR-Unsafe* — Nearly self-explanatory, this category is reserved for objects that are significantly ferromagnetic and pose a clear and direct threat to persons and equipment within the magnet room.

The very high strength of the magnetic field can also cause "missile-effect" accidents, where ferromagnetic objects are attracted to the center of the magnet, and there have been incidences of injury and death. To reduce the risks of projectile accidents, ferromagnetic objects and devices are typically prohibited in proximity to the MRI scanner and patients undergoing MRI examinations are required to remove all metallic objects, often by changing into a gown or scrubs and ferromagnetic detection devices are used by some sites There is no evidence for biological harm from even very powerful static magnetic fields

Radio Frequency Energy

A powerful radio transmitter is needed for excitation of proton spins. This can heat the body to the point of risk of hyperthermia in patients, particularly in obese patients or those with thermoregulation disorders. Several countries have issued restrictions on the maximum specific absorption rate that a scanner may produce.

Peripheral Nerve Stimulation (PNS)

The rapid switching on and off of the magnetic field gradients is capable of causing nerve stimulation. Volunteers report a twitching sensation when exposed to rapidly switched fields, particularly in their extremities. The reason the peripheral nerves are stimulated is that the changing field increases with distance from the center of the gradient coils (which more or less coincides with the center of the magnet). Note however that when imaging the head, the heart is far off-center and induction of even a tiny current into the heart must be avoided at all costs. Although PNS was not a problem for the slow, weak gradients used in the early days of MRI, the strong, rapidly switched gradients used in techniques such as EPI, fMRI, diffusion MRI, etc. are indeed capable of inducing PNS. American and European regulatory agencies insist that manufacturers stay below specified *d*B/*dt* limits (*d*B/*dt* is the change in field per unit time) or else prove that no PNS is induced for any imaging sequence. As a result of *d*B/*dt* limitation, commercial MRI systems cannot use the full rated power of their gradient amplifiers.

Acoustc Noise

Switching of field gradients causes a change in the Lorentz force experienced by the gradient coils, producing minute expansions and contractions of the coil itself. As the switching is typically in the audible frequency range, the resulting vibration produces loud noises (clicking or beeping). This is most marked with high-field machines and rapid-imaging techniques in which sound intensity can reach 120 dB(A) (equivalent to a jet engine at take-off), and therefore appropriate ear protection is essential for anyone inside the MRI scanner room during the examination.

Cryogens

As described in Physics of Magnetic Resonance Imaging, many MRI scanners rely on cryogenic liquids to enable superconducting capabilities of the electromagnetic coils within. Though the cryogenic liquids used are non-toxic, their physical properties present specific hazards. An unintentional shut-down of a superconducting electromagnet, an event known as "quench", involves the rapid boiling of liquid helium from the device. If the rapidly expanding helium cannot be dissipated through an external vent, sometimes referred to as 'quench pipe', it may be released into the scanner room where it may cause displacement of the oxygen and present a risk of asphyxiation. Oxygen deficiency monitors are usually used as a safety precaution. Liquid helium, the most commonly used cryogen in MRI, undergoes near explosive expansion as it changes from liquid to a gaseous state. The use of an Oxygen monitor is important for two reasons: 1. Keeping the O2 levels safe for patient/physicians and; 2. Making sure the MRI Scanner is working properly. Rooms built in support of superconducting MRI equipment should be equipped with pressure relief mechanisms and an exhaust fan, in addition to the required quench pipe. Since a quench results in rapid loss of all cryogens in the magnet, recommissioning the magnet is expensive and time-consuming. Spontaneous quenches are uncommon, but may also be triggered by equipment malfunction, improper cryogen fill technique, contaminants inside the cryostat, or extreme magnetic or vibrational disturbances.

Contrast Agents

The most commonly used intravenous contrast agents are based on chelates of gadolinium. In general, these agents have proved safer than the iodinated contrast agents used in X-ray radiography or CT. Anaphylactoid reactions are rare, occurring in approx. 0.03–0.1%. Of particular interest is the lower incidence of nephrotoxicity, compared with iodinated agents, when given at usual doses—this has made contrast-enhanced MRI scanning an option for patients with renal impairment, who would otherwise not be able to undergo contrast-enhanced CT. Although gadolinium agents have proved useful for patients with renal impairment, in patients with severe renal failure requiring dialysis there is a risk of a rare but serious illness, nephrogenic systemic fibrosis, that may be linked to the use of certain gadolinium-containing agents. The most frequently linked is gadodiamide, but other agents have been linked too. Although a causal link has not been definitively established, current guidelines in the United States are that dialysis patients should only receive gadolinium agents where essential, and that dialysis should be performed as soon as possible after the scan to remove the agent from the body promptly. In Europe, where more gadolinium-containing agents are available, a classification of agents according to potential risks has been released. Recently a new contrast agent named gadoxetate, brand name Eovist (US) or Primovist (EU), was approved for diagnostic use: this has the theoretical benefit of a dual excretion path.

Pregnancy

No effects of MRI on the fetus have been demonstrated. In particular, MRI avoids the use of ionizing radiation, to which the fetus is particularly sensitive. However, as a precaution, current guidelines recommend that pregnant women undergo MRI only when essential. This is particularly the case during the first trimester of pregnancy, as organogenesis takes place during this period. The concerns in pregnancy are the same as for MRI in general, but the fetus may be more sensitive to the effects—particularly to heating and to noise. However, one additional concern is the use of contrast agents; gadolinium compounds are known to cross the placenta and enter the fetal bloodstream, and it is recommended that their use be avoided. Despite these concerns, MRI is rapidly growing in importance as a way of diagnosing and monitoring congenital defects of the fetus because it can provide more diagnostic information than ultrasound and it lacks the ionizing radiation of CT. MRI without contrast agents is the imaging mode of choice for pre-surgical, in-utero diagnosis and evaluation of fetal tumors, primarily teratomas, facilitating open fetal surgery, other fetal interventions, and planning for procedures (such as the EXIT procedure) to safely deliver and treat babies whose defects would otherwise be fatal.

Claustrophobia and Discomfort

Due to the construction of some MRI scanners, they can be potentially unpleasant to lie in. Older models of closed bore MRI systems feature a fairly long tube or tunnel. The part of the body being imaged must lie at the center of the magnet, which is at the absolute center of the tunnel. Because scan times on these older scanners may be long (occasionally up to 40 minutes for the entire procedure), people with even mild claustrophobia are sometimes unable to tolerate an MRI scan without management. Modern scanners may have larger bores (up to 70 cm) and scan times are shorter. This means that claustrophobia is less of an issue, and many patients now find MRI an innocuous and easily tolerated procedure. Nervous patients may still find the following strategies helpful:

- Advance preparation
  - visiting the scanner to see the room and practice lying on the table
  - visualization techniques
  - chemical sedation
  - general anesthesia
- Coping while inside the scanner
  - holding a "panic button"
  - closing eyes as well as covering them (e.g. washcloth, eye mask)
  - listening to music on headphones or watching a movie, using mirror-glasses and a projection screen or via a Head-mounted display, while in the machine.

Alternative scanner designs, such as open or upright systems, can aso be helpful where these are available. Though open scanners have increased in popularity, they produce inferior scan quality because they operate at lower magnetic fields than closed scanners. However, commercial 1.5 tesla open systems have recently become available, providing much better image quality than previous lower field strength open models. For babies and young children chemical sedation or general anesthesia are the norm, as these subjects cannot be instructed to hold still during the scanning session. Obese patients and pregnant women may find the MRI machine to be a tight fit. Pregnant women may also have difficulty lying on their backs for an hour or more without moving.

Guidance

Safety issues, including the potential for biostimulation device interference, movement of ferromagnetic bodies, and incidental localized heating, have been addressed in the American College of Radiology's *White Paper on MR Safety,* which was originally published in 2002 and expanded in 2004. The *ACR White Paper on MR Safety* has been rewritten and was released early in 2007 under the new title *ACR Guidance Document for Safe MR Practices.* In December 2007, the Medicines in Healthcare product Regulation Agency (MHRA), a UK healthcare regulatory body, issued their *Safety Guidelines for Magnetic Resonance Imaging Equipment in Clinical Use.* In February 2008, the Joint Commission, a US healthcare accrediting organization, issued a Sentinel Event Alert #38, their highest patient safety advisory, on MRI safety issues. In July 2008, the United States Veterans Administration, a federal governmental agency serving the healthcare needs of former military personnel, issued a substantial revision to their *MRI Design Guide*, which includes physical or facility safety considerations.

The European Physical Agents Directive

The European Physical Agents (Electromagnetic Fields) Directive is legislation adopted in European legislature. Originally scheduled to be required by the end of 2008, each individual state within the European Union must include this directive in its own law by the end of 2012. Some member nations passed complying legislation and are now attempting to repeal their state laws in expectation that the final version of the EU Physical Agents Directive will be substantially revised prior to the revised adoption date. The directive applies to occupational xposure to electromagnetic fields (not medical exposure) and was intended to limit workers' acute exposure to strong electromagnetic fields, as may be found near electricity substations, radio or television transmitters or industrial equipment. However, the regulations impact significantly on MRI, with separate sections of the regulations limiting exposure to static magnetic fields, changing magnetic fields and radio frequency energy. Field strength limits are given, which may not be exceeded. An employer may commit a criminal offense by allowing a worker to exceed an exposure limit, if that is how the Directive is implemented in a particular member state. The Directive is based on the international consensus of established effects of exposure to electromagnetic fields, and in particular the advice of the European Commissions's advisor, the International Commission on Non-Ionizing Radiation Protection (ICNIRP). The aims of the Directive, and the ICNIRP guidelines it is based on, are to prevent exposure to potentially harmful fields. The actual limits in the Directive are very similar to the limits advised by the Institute of Electrical and Electronics Engineers, with the exception of the frequencies produced by the gradient coils, where the IEEE limits are significantly higher.

Many Member States of the EU already have either specific EMF regulations or (as in the UK) a general requirement under workplace health and safety legislation to protect workers against electromagnetic fields. In almost all cases the existing regulations are aligned with the ICNIRP limits so that the Directive should, in theory, have little impact on any employer already meeting their legal responsibilities. The introduction of the Directive has brought to light an existing potential issue with occupational exposures to MRI fields. There are at present very few data on the number or types of MRI practice that might lead to exposures in excess of the levels of the Directive. There is a justifiable concern amongst MRI practitioners that if the Directive were to be enforced more vigorously than existing legislation, the use of MRI might be restricted, or working practices of MRI personnel might have to change. In the initial draft a limit of static field strength to 2 T was given. This has since been removed from the regulations, and whilst it is unlikely to be restored as it was without a strong justification, some restriction on static fields may be reintroduced after the matter has been considered more fully by ICNIRP. The effect of such a limit might be to restrict the installation, operation and maintenance of MRI scanners with magnets of 2 T and stronger. As the increase in field strength has been instrumental in developing higher resolution and higher performance scanners, this would be a significant step back. This is why it is unlikely to happen without strong justification. Individual government agencies and the European Commission have now formed a working group to examine the implications on MRI and to try to address the issue of occupational exposures to electromagnetic fields from MRI.

Three-Dimensional (3D) Image Reconstruction

The Principle

Because contemporary MRI scanners offer isotropic, or near isotropic, resolution, display of images does not need to be restricted to the conventional axial images. Instead, it is possible for a software programme to build a volume by 'stacking' the individual slices one on top of the other. The programme may then display the volume in an alternative manner.

3D Rendering Techniques

Surface rendering; A threshold value of greyscale density is chosen by the operator (e.g. a level that corresponds to fat). A threshold level is set, using edge detection image processing algorithms. From this, a 3-dimensional model can be constructed and displayed on screen. Multiple models can be constructed from various different thresholds, allowing different colors to represent each anatomical component such as bone, muscle, and cartilage. However, the interior structure of each element is not visible in this mode of operation.

Volume rendering: Surface rendering is limited in that it only displays surfaces that meet a threshold density, and only displays the surface closest to the imaginary viewer. In volume rendering, transparency and colors are used to allow a better representation of the volume to be shown in a single image - e.g. the bones of the pelvis could be displayed as semi-transparent, so that even at an oblique angle, one part of the image does not conceal another.

Image Segmentation

Where different structures have similar threshold density, it can become impossible to separate them simply by adjusting volume rendering parameters. The solution is called segmentation, a manual or automatic procedure that can remove the unwanted structures from the image.

2003 Nobel Prize

Reflecting the fundamental importance and applicability of MRI in medicine, Paul Lauterbur of the University of Illinois at Urbana-Champaign and Sir Peter Mansfield of the University of Nottingham were awarded the 2003 Nobel Prize in Physiology or Medicine for their "discoveries concerning magnetic resonance imaging". The Nobel citation acknowledged Lauterbur's insight of using magnetic field gradients to determine spatial localization, a discovery that allowed rapid acquisition of 2D images. Mansfield was credited with introducing the mathematical formalism and developing techniques for efficient gradient utilization and fast imaging. The actual research that won the prize was done almost 30 years before, while Paul Lauterbur was at Stony Brook University in New York. The award was vigorously protested by Raymond Vahan Damadian, founder of FONAR Corporation, who claimed that he invented the MRI, and that Lauterbur and Mansfield had merely refined the technology. An ad hoc group, called "The Friends of Raymond Damadian", took out full-page advertisements in the *New York Times* and *The Washington Post* entitled "The Shameful Wrong That Must Be Righted", demanding that he be awarded at least a share of the Nobel Prize. Also, even earlier, in the Soviet Union, Vladislav Ivanov filed (in 1960) a document with the USSR State Committee for Inventions and Discovery at Leningrad for a Magnetic Resonance Imaging device, although this was not approved until the 1970s. In a letter to *Physics Today*, Herman Carr pointed out his own even earlier use of field gradients for one-dimensional MR imaging.

Virtual Particle

In physics, a virtual particle is a particle that exists for a limited time and space, introducing uncertainty in its energy and momentum due to the uncertainty principle. Because energy and momentum in quantum mechanics are time and space derivative operators, then due to Fourier transforms their spans are inversely proportional to time duration and position spans, respectively Virtual particles exhibit some of the phenomena that real particles do, such as obedience to the conservation laws.

If a single particle is detected, then the consequences of its existence are prolonged to such a degree that it cannot be virtual. Virtual particles are viewed as the quanta that describe fields of the basic force interactions, which cannot be described in terms of real particles. Examples of these are static force fields, such as a simple electric or magnetic field, or any field that exists without excitations that result in its carrying informtion from place to place. Virtual photons are also a major component of antenna near field phenomena and induction fields, which have only very short-range effects, that do not radiate through space with the same range-properties as do electromagnetic wave photons. For example, the energy carried from one winding of a transformer to another, in quantum terms, is carried by virtual photons, not real photons. The virtual particle forms of massless particles, such as photons, do have mass (which may be either positive or negative) and are said to be off mass shell. They are allowed to have mass (which consists of "borrowed energy") because they exist for only a temporary time, which in turn gives them a limited "range". This is in accordance with the uncertainty principle which allows existence of such particles of borrowed energy, so long as their energy, multiplied by the time they exist, is a fraction of Planck's constant. The concept of virtual particles is closely related to the idea of quantum fluctuations. Virtual particles can be thought of as coming into existence as quantities, such as the electric field, fluctuate around their expectation values as required by quantum mechanics.

Properties

The concept of virtual particles arises in the perturbation theory of quantum field theory, an approximation scheme in which interactions (essentially forces) between real particles are calculated in terms of exchanges of virtual particles. Any process involving virtual particles admits a schematic representation known as a Feynman diagram which facilitates the understanding of calculations. A virtual particle is one that does not precisely obey the $m^2c^4 = E^2 - p^2c^2$ relationship for a short time. In other words, its kinetic energy may not have the usual relationship to velocity–indeed, it can be negative.

The probability amplitude for it to exist tends to be canceled out by destructive interference over longer distances and times. A virtual particle can be considered a manifestation of quantum tunnelling. The range of forces carried by virtual particles is limited by the uncertainty principle, which regards energy and time as conjugate variables; thus virtual particles of larger mass have more limited range. There is not a definite line differentiating virtual particles from real particles — the equations of physics just describe particles (which includes both equally). The amplitude that a virtual particle exists interferes with the amplitude for its non-existence; whereas for a real particle the cases of existence and non-existence cease to be coherent with each other and do not interfere any more. In the quantum field theory view, "real particles" are viewed as being detectable excitations of underlying quantum fields. As such, virtual particles are also excitations of the underlying fields, but are detectable only as forces but not particles. They are "temporary" in the sense that they appear in calculations, but are not detected as single particles. Thus, in mathematical terms, they never appear as indices to the scattering matrix, which is to say, they never appear as the observable inputs and outputs of the physical process being modelled. In this sense, virtual particles are an artifact of perturbation theory, and do not appear in a non-perturbative treatment. There are two principal ways in which the notion of virtual particles appears in modern physics. They appear as intermediate terms in Feynman diagrams; that is, as terms in a perturbative calculation. They also appear as an infinite set of states to be summed or integrated over in the calculation of a semi-non-perturbative effect. In the latter case, it is sometimes said that virtual particles cause the effect, or that the effect occurs because of the existence of virtual particles.

Manifestations

There are many observable physical phenomena resulting from interactions involving virtual particles. For bosonic particles which exhibit rest mass when they are free and "real," virtual interactions are characterized by the relatively short range of the force interaction produced by particle exchange. Examples of such short-range interactions are the strong and weak forces, and their associated field bosons. For the gravitational and electromagnetic forces, the zero rest-mass of the associated boson particle permits long-range forces to be mediated by virtual particles. However, in the case of photons, power and information transfer by virtual particles is a relatively short-range phenomenon (existing importantly only within a few wavelengths of the field-disturbance which carries information or transferred power), as for example seen in the characteristically short range of inductive and capacitative effects in the near field zone of coils and antennas. Some field interactions which may be seen in terms of virtual particles are:

- The Coulomb force (static electric force) between electric charges. It is caused by the exchange of virtual photons. In symmetric 3-dimensional space this exchange results in the inverse square law for electric force. Since the photon has no mass, the coulomb potential has an infinite range.
- The magnetic field between magnetic dipoles. It is caused by the exchange of virtual photons. In symmetric 3-dimensional

space this exchange results in the inverse square law for magnetic force. Since the photon has no mass, the magnetic potential has an infinite range.

- The so-called near field of radio antennas, where the magnetic and electric effects of the changing current in the antenna wire and the charge effects of the wire's capacitive charge are detectable, but both of which effects decay with increasing distance from the antenna much more quickly than do the influence of conventional electromagnetic waves, for which E is always equal to cB, and which are composed of real photons.
- The strong nuclear force between quarks is the result of interaction of virtual gluons. The residual of this force outside of quark triplets (neutron and proton) holds neutrons and protons together in nuclei, and is due to virtual mesons such as the pi meson and rho meson.
- The weak nuclear force - it is the result of exchange by virtual W and Z bosons.
- The spontaneous emission of a photon during the decay of an excited atom or excited nucleus; such a decay is prohibited by ordinary quantum mechanics and requires the quantization of the electromagnetic field for its explanation.
- The Casimir effect, where the ground state of the quantized electromagnetic field causes attraction between a pair of electrically neutral metal plates.
- The van der Waals force, which is partly due to the Casimir effect between two atoms,
- Vacuum polarization, which involves pair production or the decay of the vacuum, which is the spontaneous production of particle-antiparticle pairs (such as electron-positron).
- Lamb shift of positions of atomic levels.
- Hawking radiation, where the gravitational field is so strong that it causes the spontaneous production of photon pairs (with black body energy distribution) and even of particle pairs.

Most of these have analogous effects in solid-state physics; indeed, one can often gain a better intuitive understanding by examining these cases. In semiconductors, the roles of electrons, positrons and photons in field theory are replaced by electrons in the conduction band, holes in the valence band, and phonons or vibrations of the crystal lattice. A virtual particle is in a virtual state where the probability amplitude is not conserved. Examples of macroscopic virtual phonons, photons, and electrons in the case of the tunneling process were presented by Günter Nimtz in.

History

Paul Dirac was the first to propose that empty space (a vacuum) can be visualized as consisting of a sea of virtual electron-positron pairs, known as the Dirac sea. The Dirac sea has a direct analog to the electronic band structure in crystalline solids as described in solid state physics. Here, particles correspond to conduction electrons, and antiparticles to holes. A variety of interesting phenomena can be attributed to this structure.

Virtual Particles in Feynman Diagrams

The calculation of scattering amplitudes in theoretical particle physics requires the use of some rather large and complicated integrals over a large number of variables. These integrals do, however, have a regular structure, and may be represented as Feynman diagrams. The appeal of the Feynman diagrams is strong, as it allows for a simple visual presentation of what would otherwise be a rather arcane and abstract formula. In particular, part of the appeal is that the outgoing legs of a Feynman diagram can be associated with real, on-shell particles. Thus, it is natural to associate the other lines in the diagram with particles as well, called the "virtual particles". Mathematically, they correspond to the propagators appearing in the diagram. In the image to the right, the solid lines correspond to real particles (of momentum p_1 and so on), while the dotted line corresponds to a virtual particle carrying momentum k. For example, if the solid lines were to correspond to electrons interacting by means of the electromagnetic interaction, the dotted line would correspond to the exchange of a virtual photon. In the case of interacting nucleons, the dotted line would be a virtual pion. In the case of quarks interacting by means of the strong force, the dotted line would be a virtual gluon, and so on.

It is sometimes said that all photons are virtual photons. This is because the world-lines of photons always resemble the dotted line in the above Feynman diagram: the photon was emitted somewhere (say, a distant star), and then is absorbed somewhere else (say a photoreceptor cell in the eyeball). Furthermore, in a vacuum, a photon experiences no passage of (proper) time between emission and absorption. This statement illustrates the difficulty of trying to distinguish between "real" and "virtual" particles as mathematically they are the same objects and it is only our definition of "reality" which is weak here. In practice, a clear distinction can be made: real photons are detected as individual particles in particle detectors, whereas virtual photons are not directly detected; only their average or side-effects may be noticed, in the form of *forces* or (in modern language) *interactions* between particles. Virtual particles need not be mesons or vector bosons, as in the example above; they may also be fermions. However, in order to preserve quantum numbers, most simple diagrams involving fermion exchange are prohibited. The image to the right shows an allowed diagram, a one-loop diagram. The solid lines correspond to a fermion propagator, the wavy lines to bosons.

Pair Production

In order to conserve the total fermion number of the universe, a fermion cannot be created without also creating its antiparticle; thus many physical processes lead to pair creation. The need for the normal ordering of particle fields in the vacuum can be interpreted by the idea that a pair of virtual particles may briefly "pop into existence", and then annihilate each other a short while later. Thus, virtual particles are often popularly described as coming in pairs, a particle and antiparticle, which can be of any kind. These pairs exist for an extremely short time, and mutually annihilate in short order. In some cases, however, it is possible to boost the pair apart using external energy so that they avoid annihilation and become real particles. This may occur in one of two ways. In an accelerating frame of reference, the virtual particles may appear to be real to the accelerating observer; this is known as the Unruh effect. In short, the vacuum of a stationary frame appears, to the accelerated observer, to be a warm gas of real particles in thermodynamic equilibrium. The Unruh effect is a toy model for understanding Hawking radiation, the process by which black holes evaporate. Another example is pair production in very strong electric fields, sometimes called vacuum decay. If, for example, a pair of atomic nuclei are merged together to very briefly form a nucleus with a charge greater than about 140, (that is, larger than about the inverse of the fine structure constant), the strength of the electric field will be such that it will be energetically favourable to create positron-electron pairs out of the vacuum or Dirac sea, with the electron attracted to the nucleus to annihilate the positive charge.

This pair-creation amplitude was first calculated by Julian Schwinger in 1951. The restriction to particle-antiparticle pairs is actually only necessary if the particles in question carry a conserved quantity, such as electric charge, which is not present in the initial or final state. Otherwise, other situations can arise. For instance, the beta decay of a neutron can happen through the emission of a single virtual, negatively charged W particle that almost immediately decays into a real electron and antineutrino; the neutron turns into a proton when it emits the W particle. The evaporation of a black hole is a process dominated by photons, which are their own antiparticles and are uncharged.

Spontaneous Fission

Spontaneous fission (SF) is a form of radioactive decay characteristic of very heavy isotopes. It is theoretically possible for any atomic nucleus whose mass is greater than or equal to 100 atomic mass units (u), i.e. elements near ruthenium. In practice, however, spontaneous fission is only energetically feasible for atomic masses above 230 u (elements near thorium). The elements most susceptible to spontaneous fission are the high-atomic-number actinide elements, such as mendelevium and lawrencium, and the trans-actinide elements, such as rutherfordium. For uranium and thorium, the spontaneous fission mode of decay does occur, but it is not seen for the majority of radioactive breakdowns, and it is usually neglected except for the exact considerations of branching ratios when determining the activity of a sample containing these elements. As the name suggests, spontaneous fission gives the same result as induced nuclear fission. However, like other forms of radioactive decay, it occurs due to quantum tunneling, without the atom having been struck by a neutron or other particle as in induced nuclear fission. Spontaneous fissions release neutrons as all fissions do, so if a critical mass is present, a spontaneous fission can initiate a chain reaction. Also, radioisotopes for which spontaneous fission is a nonnegligible decay mode may be used as neutron sources; californium-252 (half-life 2.645 years, SF branch ratio 3.09%) is often used for this purpose.

The neutrons may then be used to inspect airline luggage for hidden explosives, to gauge the moisture content of soil in the road construction and building industries, to measure the moisture of materials stored in silos, and in other applications. As long as the fissions give a negligible reduction of the amount of nuclei that can spontaneously fission, this is a Poisson process: for very short time intervals the probability of a spontaneous fission is proportional to the length of time. The spontaneous fission of uranium-238 leaves trails of damage in uranium-bearing minerals as the fission fragments recoil through the crystal structure. These trails, or *fission tracks*, provide the basis for the radiometric dating technique known as fission track dating.

History

The first nuclear fission process discovered was the fission induced by neutrons. Because cosmic rays produce some neutrons, it was difficult to distinguish between induced and spontaneous fission events. Cosmic rays can be reliably shielded by a thick layer of rock or water. The spontaneous fission was identified in 1940 by Soviet physicists Georgy Flyorov and Konstantin Petrzhak by their observations of uranium in the Moscow Metro Dinamo station, 60 metres (200 ft) deep underground.

Neutron Emission

Neutron emission is a type of radioactive decay of atoms containing excess neutrons, in which a neutron is simply ejected from the nucleus. Two examples of isotopes which emit neutrons are helium-5 and beryllium-13. However, the decay of helium-5 is also (by definition) a case of alpha-decay. Many heavy isotopes, most notably californium-252, emit neutrons among the products of a different radioactive decay process, spontaneous fission. Neutrons are absorbed and emitted in the process of nuclear fission, a nuclear chain reaction propagated by neutrons. Delayed neutrons emitted by neutron-rich fission products aid control of nuclear reactors by making reactivity change much slower than it would be for prompt neutrons alone.

Positron Emission

Positron emission or beta plus decay (β^+ decay) is a type of beta decay in which a proton is converted, via the weak force, to a neutron, releasing a positron (the antimatter counterpart of an electron) and a neutrino. Isotopes which undergo this decay and thereby emit positrons include carbon-11, potassium-40, nitrogen-13, oxygen-15, fluorine-18, and iodine-121. Inside protons and neutrons, there are fundamental particles called quarks. The two most common types of quarks are up quarks, which have a charge of $+^2/_3$ and down quarks, with a $-^1/_3$ charge. Quarks arrange themselves in sets of three such that they make protons and neutrons. In a proton, whose charge is +1, there are two up quarks and one down quark. Neutrons, with no charge, have one up quark and two down quarks. Quarks are able to change from up quarks to down quarks.

It is this that causes beta radiation. Positron emission happens when an up quark changes into a down quark. These isotopes are used in positron emission tomography, a technique used for medical imaging. Note that the energy emitted depends on the isotope that is decaying; the figure of 0.96 MeV applies only to the decay of carbon-11. Isotopes which increase in mass under the conversion of a proton to a neutron, or which decrease by less than $2m_e$, do not spontaneously decay by positron emission. Nuclei which decay by positron emission may also decay by electron capture. For low-energy decays, electron capture is energetically favoured by $2m_ec^2 = 1.022$ MeV, since the final state has an electron removed rather than a positron added. As the energy of the decay goes up, so does the branching ratio towards positron emission. However, if the energy difference is less than $2m_ec^2$, then positron emission cannot occur and electron capture is the sole decay mode. Certain isotopes (for instance, ^{7}Be) are stable in galactic cosmic rays, because the electrons are stripped away and the decay energy is too small for positron emission.

Proton Emission

Proton emission (also known as proton radioactivity) is a type of radioactive decay in which a proton is ejected from a nucleus. Proton emission can occur from high-lying excited states in a nucleus following a beta decay, in which case the process is known as beta-delayed proton emission, or can occur from the ground state (or a low-lying isomer) of very proton-rich nuclei, in which case the process is very similar to alpha decay. For a proton to escape a nucleus, the proton separation energy must be negative - the proton is therefore unbound, and tunnels out of the nucleus in a finite time. Proton emission is not seen in naturally-occurring isotopes; proton emitters can be produced via nuclear reactions, usually utilising some kind of particle accelerator. Although prompt (i.e. not beta-delayed) proton emission was observed from an isomer in Cobalt-53 as early as 1969, no other proton-emitting states were found until 1981, when the proton radioactive ground states of Lutetium-151 and Thulium-147 were observed at experiments at the GSI in West Germany. Research in the field flourished after this breakthrough, and to date more than 25 isotopes have been found to exhibit proton emission. The study of proton emission has aided the understanding of nuclear deformation, masses and structure, and it is a wonderfully pure example of quantum tunneling. In 2002, the simultaneous emission of two protons was observed from the nucleus Iron-45 in experiments at GSI and GANIL (Grand Accélérateur National d'Ions Lourds, near Caen). In 2005 it was experimentally determined (at the same facility) that Zinc-54 can also undergo double proton decay.

Electron Capture

Electron capture is a process in which a proton-rich nuclide absorbs an inner atomic electron (changing a nuclear proton to a neutron) and simultaneously emits a neutrino. Various photon emissions follow, in order to allow the energy of the atom to fall to the ground state of the new nuclide. Electron capture is the primary decay mode for isotopes that have a relative superabundance of protons in the nucleus, but there is insufficient energy difference between the isotope and its prospective daughter with one less positive charge, for the nuclide to decay by simply emitting a positron.

Electron capture also exists as a viable decay mode for radioactive isotopes that do have enough energy to decay by positron emission, and, in that case, it competes with positron emission. It is sometimes called inverse beta decay, though this term can also refer to the capture of a neutrino through a similar process. If the energy difference between the parent atom and the daughter atom is less than 1.022 MeV, positron emission is forbidden because not enough decay energy is available to allow it, and thus electron capture is the sole decay mode. For example, rubidium-83 (37 protons, 46 neutrons) will decay to krypton-83 (36 protons, 47 neutrons) solely by electron capture (the energy difference, or decay energy, is about 0.9 MeV).

Since the proton is changed to a neutron in electron capture, the number of neutrons increases by 1, the number of protons decreases by 1, and the atomic mass number remains unchanged. By changing the number of protons, electron capture transforms the nuclide into a new element. The atom, although still neutral in charge, now exists in an energetically excited state with the inner shell missing an electron. While transiting to the ground state, the atom will emit an X-ray photon (a type of electromagnetic radiation) and/or Auger electrons, or both. Often the nucleus exists in an excited state as well, and emits a gamma ray in order to reach the ground state energy of the new nuclide just formed.

History

The theory of electron capture was first discussed by Gian-Carlo Wick in a 1934 paper, and then developed by Hideki Yukawa and others. K-electron capture was first observed by Luis Alvarez, in vanadium-48. He reported it in a 1937 paper in the *Physical Review*. Alvarez went on to study electron capture in gallium-67 and other nuclides. Note that it is one of the initial atom's own electrons that is captured, not a new, incoming electron, as might be suggested by the way the above reactions are written. Radioactive isotopes that decay by pure electron capture can, in theory, be inhibited from radioactive decay if they are fully ionized ("stripped" is sometimes used to describe such ions). It is hypothesized that such elements, if formed by the r-process in exploding supernovae, are ejected fully ionized and so do not undergo radioactive decay as long as they do not encounter electrons in outer space. Anomalies in elemental distributions are thought to be partly a result of this effect on electron capture. Chemical bonds can also affect the rate of electron capture to a small degree (in general, less than 1%) depending on the proximity of electrons to the nucleus. For example in ^{7}Be, a difference of 0.9% has been observed between half-lives in metallic and insulating environments. This relatively large effect is due to the fact that beryllium is a small atom whose valence electrons are close to the nucleus. Around the elements in the middle of the periodic table, isotopes that are lighter than stable isotopes of the same element tend to decay through electron capture, while isotopes heavier than the stable ones decay by electron emission.

Neutron Capture

Neutron capture is a kind of nuclear reaction in which an atomic nucleus collides with one or more neutrons and they merge to form a heavier nucleus. Since neutrons have no electric charge they can enter a nucleus more easily than positively charged particles, which are repelled electrostatically. Neutron capture plays an important role in the cosmic nucleosynthesis of heavy elements. In stars it can proceed in two ways - as a rapid process (an r-process) or a slow process (an s-process). Nuclei of masses greater than 56 cannot be formed by thermonuclear reactions (i.e. by nuclear fusion), but can be formed by neutron capture.

Neutron Capture at Small Neutron Flux

At small neutron flux, as in a nuclear reactor, a single neutron is captured by a nucleus. For example when natural gold (^{197}Au) is irradiated by neutrons the isotope ^{198}Au is formed in a highly excited state which then quickly decays to the ground state of ^{198}Au by the emission of γ rays. In this process the mass number (the number of nucleons - both protons and neutrons) increases by one. In terms of a formula this is written $^{197}Au(n,\gamma)^{198}Au$. If thermal neutrons are used we call it thermal capture. The isotope ^{198}Au is a beta emitter that decays into the mercury isotope ^{198}Hg. In this process the atomic number (the number of protons in the nucleus) rises by one. The s-process mentioned above happens in the same way, but inside of stars.

Neutron Capture at High Neutron Flux

The r-process happens inside stars if the neutron flux density is so high that the atomic nucleus has no time to decay via beta emission in between neutron captures. The mass number therefore rises by a large amount while the atomic number (i.e., the element) stays the same. Only afterwards, the highly unstable nuclei decay via many β^- decays to stable or unstable nuclei of high atomic number.

Capture Cross Section

The absorption neutron cross-section of an isotope of a chemical element is the effective cross sectional area that an atom of that isotope presents to absorption, and is a measure of the probability of neutron capture. It is usually measured in barns (b). Absorption cross section is often highly dependent on neutron energy. Two of the most commonly specified measures are the cross-section for thermal neutron absorption, and resonance integral which considers the contribution of absorption peaks at certain neutron energies specific to a particular nuclide, usually above the thermal range, but encountered as neutron moderation slows the neutron down from an original high energy. The thermal energy of the nucleus also has an effect; as temperatures rise, Doppler broadening increases the chance of catching a resonance peak. In particular, the increase in U-238's ability to absorb neutrons at higher temperatures (and to do so without fissioning) is a negative feedback mechanism that helps keep nuclear reactors under control.

Uses

Neutron activation analysis can be used to remotely detect the chemical composition of materials. This is because different elements release different characteristic radiation when they absorb neutrons. This makes t useful in many fields related to mineral exploration and security.

Neutron Absorbers

The most prolific neutron absorbers are the radioactive isotopes of elements that happen to become stable by absorbing one neutron. An example of these is xenon-135 (half life about 9.1 hours), which absorbs a neutron to become the stable isotope xenon-136. Xenon-135 is formed in nuclear reactors when the splitting of uranium-235, uranium-233, or plutonium-239, in a nuclear chain reaction commonly leads to the production of some iodine-135. Iodine-135 soon undergoes nuclear decay, by emitting a beta particle - with a quite short half-life - to produce xenon-135. Other isotopes that are major neutron-absorbers include helium-3, which becomes tritium, a heavy isotope of hydrogen; and boron-10, which splits to become lithium-7 and a helium-4 nucleus. Samarium-149, which is also formed during nuclear chain reactions, is also a highly effective neutron absorber. It also has a very long half-life, and it lasts effectively permanently in the nuclear fuel until it absorbs a neutron, which transmutes it into samarium-150, a stable isotope. Other important neutron absorbers that are used in nuclear reactors for their control rods include the metals cadmium, hafnium, and sometimes the rare earth element, gadolinium, all of which usually consist of mixtures of various isotopes—some of which are excellent neutron-absorbers. Hafnium, one of the last stable elements to be discovered, presents an interesting case. Even though hafnium is a heaver element, its electron configuration makes it practically identical with the element zirconium, and they are always found in the same ores, and in the gem, the zircon. However, their nuclear properties are different in a profound way. Hafnium absorbs neutrons avidly, and it can be used in reactor control rods. However, natural zirconium is practically transparent to neutrons. Thus, zirconium is very desirable for use in the construction of reactors, including in such parts as the metal cladding of the fuel rods which contain either uranium, plutonium, or an alloy of the two metals. Hence, it is quite important to be able to separate the zirconium from the hafnium in their naturally-occurring alloy. This can only be done inexpensively by using modern chemical ion-exchange resins.

Bibliography

Antsiferov, V. V. and G. I. Smirnov: *Coherent Radiation Processes in Plasmas*, Cambridge Univ. Press, Cambridge, 1999.

Balakrishan, A.: *Biofuel Crops and Renewable Energy*, Sublime Publications, Delhi, 2011.

Bard, A.J.; Faulkner, L.R.: *Electrochemical Methods: Fundamentals and Applications*, John Wiley & Sons, New York, 2005.

Basdevant, J.L., and J. Dalibard: *Quantum Mechanics*, Springer, Berlin, 2005.

Benenson, W., J. Harris, H. Stoecker, , and H. Lutz: *Handbook of Physics*, Springer, Berlin, 2006.

Braams, C. and P. Stott: *Nuclear Fusion: Half a Century of Magnetic Confinement Fusion*, IOP Press, NJ, 2002.

Brauman J. I. and Coulson C. A.: *Supplementary Tables of Molecular Orbital Calculations*, Pergamon Press, Oxford, 1965.

Brush, S.: *The Kind of Motion That We Call Heat*, North-Holland, Amsterdam, 1976.

Buchner, J.: *Physics of Space Plasmas*, MIT Press, US, 1995.

Buckel, W.: *Superconductivity: Fundamentals and Applications,* VCH Publishers, Inc., 1991.

Busch, P., P. Lahti, and P. Mittelstaedt: *The Quantum Theory of Measurement*, Springer-Verlag, Berlin, 1991.

Callen, H. B.: *Thermodynamics*, Wiley, New York, 1960.

Charles F. Kennel, *Convection and Substorms*, Oxford Univ. Press, Oxford, 1995.

Cox, Brian; Forshaw, Jeff: *The Quantum Universe: Everything That Can Happen Does Happen*, Allen Lane, UK, 2011.

Cross, R. C.: *An Introduction to Alfven Waves*, Hilger, Bristol, 1988.

d'Espagnat, B.: *Veiled Reality*, Addison-Wesley, Reading, MA, 1995.

Dahl, P. F.: *Superconductivity: Its Historical Roots and Development from Mercury to the Ceramic Oxides,* American Institute of Physics, 1992.

Daniel Hastings, Henry Garrett: *Spacecraft-Environment Interactions*, Cambridge University Press, Cambridge, 2000.

Das, A. C.: *Space Plasma Physics: An Introduction,* Narosa Publishing House, New Delhi, 2004.

David Chandler: *Introduction to Modern Statistical Mechanics,* Oxford University Press, Oxford, 1987.

Davidson, R.: *The Physics of Nonneutral Plasmas*, Imperial College Press, London, 2001.

Davidson, Ronald: *An Introduction to the Physics of Nonneutral Plasma*, Addison-Wesley, London, 1990.

Donald A. McQuarrie: *Statistical Mechanics,* Harper and Row, UK, 1976.

Dunai, T.J.: *Cosmogenic Nucleides,* Cambridge University Press, U.K., 2010.

Ehrenfest, P. and T.: *The Conceptual Foundations of the Statistical Approach in Mechanics*, Cornell University Press, Ithaca, NY, 1959.

Freidberg, J. P.: *Ideal Magnetohydrodynamics*, Plenum Pr., NY, 1987.

Freyd, P.: *Functor Theory,* Princeton University, Princeton, New Jersey, 1960.

Gabriel, P. and M.Zisman:*Category of Fractions and Homotopy Theory*, Springer, Berlin, 1967.

Gareth Wynn-Williams: *The Fullness of Space,* Cambridge Press, Cambridge, 1992.

Gasiorowicz, S.: *Quantum Physics,* Wiley, New York, 1995.

George J.: *Fuel Cells,* New York: Reinhold Publishing Corp., 1961.

Gerd W. Prolss: *Physics of the Earth's Space Environment,* Springer-Verlag, London, 2004.

Ghilardi, S. & Zawadowski, M.: *Model Completions: A Categorical Approach to Nonclassical Porpositional Logics*, Dordrecht, Kluwer, 2002.

Gibbon, Paul: *Short Pulse Laser Interactions with Matter*, Imperial College Press, London, 2000.

Gibbs, J.: *Elementary Principles in Statistical Mechanics*, Dover, New York, 1960.

Giuliano Benenti: *Principles of Quantum Computation and Information,* World Scientific, New Jersey, 2004.

Goedbloed, J. and S. Poedts: *Principles of Magnetohydrodynamics: With Applications to Laboratory and Astrophysical Plasmas,* Cambridge Press, Cambridge, 2004.

Gorse, C.: *Plasma Technology: Fundamentals and Applications*, Plenum Press, NY, 1992.

Griffiths, D.: *Introduction to Quantum Mechanics,* Prentice Hall, Upper Saddle River, NJ, 1995.

Gunther Rudiger and Rainer Hollerbach: *The Magnetic Universe: Geophysical and Astrophysical Dynamo Theory* , Wiley-VCH, 2004.

Halmos, P.: *Introduction to Hilbert Space,* AMS Chelsea Publishing, Providence, 1957.

Harms, A. A.: *Principles of Fusion Energy,* World Scientific, NJ, 2000.

Hatcher, W. S.: *The Logical Foundations of Mathematics*, Pergamon Press, Oxford, 1982.

Hill, Terrell L.: *Introduction to Statistical Thermodynamics*, Dover Publications, New York, 1986.

Ibach H. and H. Lýÿth: *Solid-State Physics: an Introduction to Principles of Materials Science,* Springer, Berlin; New York, 1996.

John W. Freeman: *Storms in Space,* Cambridge, 2001.

Johnstone, P. T.: *Topos Theory*, Academic Press, New York, 1977.

Kallenrode, M.B.: *Space Physics: Plasmas and particles in the Heliosphere and Magnetosheres*, Springer, London, 1998.

Kenneth, T. Fowler: *The Fusion Quest,* Johns Hopkins Press, Baltimore, MD, 1997.

Kresin, V. Z. and S. A. Wolf: *Fundamentals of Superconductivity,* Plenum Press, New York, 1990.

Lambek, J. and P. J. Scott: *Introduction to Higher Order Categorical Logic*, Cambridge University Press, Cambridge, 1986.

Leonard F. Burlaga: *Interplanetary Magnetohydrodynamics*, Oxford Univ. Press, Oxford, 1995.

Lequeux, James: *The Interstellar Medium,* Springer-Verlag, London, 2005.

Lindl, J.: *Inertial Confinement Fusion*, Springer, NJ, 1998.

Mac Lane, S.: *Mathematics, Form and Function*, Springer, New York, 1986.

Mark W. Zemansky: *Heat and Thermodynamics*, McGraw-Hill, Delhi, 1968.

McCracken, G. and P. Stott: *Fusion: The Energy of the Universe*, Elsevier, London, 2005.

Mehra, J. and H. Rechenberg: *The Historical Development of Quantum Theory*, Springer-Verlag, New York 1982.

Merzbacher, E.: *Quantum Mechanics,* Wiley, New York, 1997.

Moiseev, S. S.; V. Oraevsky, and V. Pungin: *Non-Linear Instabilities in Plasmas and Hydrodynamics*, IOP Press, NJ, 1999.

Neil Gershenfeld: *The Physics of Information Technology,* Cambridge Univ. Press, Cambridge, 2000.

Pareigis, B.: *Categories and Functors*, Academic Press, New York, 1970.

Planck, M.: *A Survey of Physical Theory*, Methuen & Co., Ltd., London, 1925.

Plymen, R.J. and P. L. Robinson: *Spinors in Hilbert Space*, Cambridge Univ. Press, Cambridge, 1994.

Popescu, N.: *Abelian Categories with Applications to Rings and Modules,* Academic Press., New York and London, 1973.

Price, H.: *Time's Arrow and the Archimedean Point*, Oxford University Press, Oxford, 1996.

Primas, H.: *Quantum Mechanics, Chemistry and Reductionism,* Springer, Berlin, 1983.

Rae, A.: *Quantum Physics: Illusion or Reality?*, Cambridge University Press, Cambridge, 1986.

Redhead, M.L.G.: *Incompleteness, Nonlocality and Realism*, Clarendon Press, Oxford, 1989.

Scott, P. J.: *Some Aspects of Categories in Computer Science*, Amsterdam, North Holland, 2000.

Shevelko, V. and L. Vainshtein: *Atomic Physics for Hot Plasmas*, Oxford Press, Oxford, 1993.

Sklar, L.: *Physics and Chance: Philosophical Issues in the Foundations of Statistical Mechanics*, Cambridge University Press, Cambridge, 1993.

Squires, E.: *Conscious Mind in the Physical World*, Adam Hilger, Bristol, New York, 1990.

Stangeby, P. C.: *The Plasma Boundary of Magnetic Fusion Devices*, IOP Press, NJ, 2000.

Stephanie Frank Singer: *Linearity, Symmetry, and Prediction in the Hydrogen Atom*, Springer, New York, 2005.

Unruh, W.G.: *Physics Meets Philosophy at the Planck Scale*, Cambridge University Press, Cambridge, 2001.

Vidali, G.: *Superconductivity: The Next Revolution?,* Cambridge University Press, New York, NY, USA, 1993.

Wakatani, M.: *Stellerator and Heliotron Devices*, Oxford Univ. Press, Oxford, 1998.

Weinberg, S.: *The Quantum Theory of Fields*, Cambridge Univ. Press, Cambridge, 1995.

Wesson, J.: *Tokomaks,* Oxford Univ. Press, Oxford, 2004.

Whitaker, A.: *Einstein, Bohr and the Quantum Dilemma*, Cambridge University Press, Cambridge, 1996.

Wielebinski, R. and R. Beck: *Cosmic Magnetic Fields*, Springer-Verlag, London, 2005.

Wilhelmsson, H. and E. Lazzaro: *Reaction-Diffusion in the Physics of Hot Plasmas*, IOP, NJ, 2000.

Wolfgang Kundt: *Astrophysics: A New Approach,* Springer-Verlag, London, 2005.

Index

N

P

Q

R

S

V

W

X

❑❑❑